Spatio-Temporal Analysis and Optimization of Land Use/Cover Change

Spatio-Temporal Analysis and Optimization of Land Use/Cover Change

Biao Liu

Center for Assessment and Development of Real Estate, Shenzhen, China; Key Laboratory of the Ministry of Land and Resources for Construction Land Transformation, Guangzhou, China; Guangdong Province Key Laboratory for Land use and consolidation, Guangzhou, China; Guangdong Province Engineering Research Center for Land Information Technology, Guangzhou, China

Bo Huang

Shenzhen Research Institute, The Chinese University of Hong Kong; Department of Geography and Resource Management and Institute of Space and Earth Information Science (ISEIS), The Chinese University of Hong Kong, Hong Kong, China

Wenting Zhang

College of resources and environment, Huazhong Agricultural University, Wuhan, China

CRC Press
Taylor & Francis Group
Boca Raton London New York

CRC Press is an imprint of the
Taylor & Francis Group, an **informa** business

A BALKEMA BOOK

Project 41371417 supported by National Natural Science Foundation of China

CRC Press
Taylor & Francis Group
6000 Broken Sound Parkway NW, Suite 300
Boca Raton, FL 33487-2742

First issued in paperback 2019

© 2017 by Taylor & Francis Group, LLC
CRC Press is an imprint of Taylor & Francis Group, an Informa business

Typeset by Integra Software Services Private Ltd

ISBN 13: 978-1-138-03315-3 (hbk)
ISBN 13: 978-0-367-88575-5 (pbk)

Library of Congress Cataloging in Publication Data

**Visit the Taylor & Francis Web site at
http://www.taylorandfrancis.com**

**and the CRC Press Web site at
http://www.crcpress.com**

Contents

Preface

On the occasion of the official publication of this book, I am very pleased to be invited to write the preface. I hope that this book's publication will not only encourage experts and scholars at home and abroad to carry out further theoretical research work on land use, but that it will also attract the attention of policy-makers, organizers and implementers of land use planning. As a result, we will jointly achieve the scientific, reasonable and efficient use of land resources.

There is no doubt that land is a scarce natural resource, and that it is significantly affected by the competition of mutually exclusive uses because all human activities require space. From a worldwide perspective, and increasingly since the nineteenth century, human activities and especially industrialization have greatly changed the land cover of the whole world; this mainly involves the reduction of forest and woodland, the pollution of rivers, and the trend of urbanization. All of these have not only already changed the earth profoundly, but also they are continuing to have a great impact on the environment. In our country, after the initial disorderly and extensive use of land resources, we are increasingly feeling a lack of such resources, especially in the cities. Is it simply because our land resources are insufficient? Of course it is not. I think that one of the important reasons for this problem is the lack of scientific land use planning in advance, so that we have failed to achieve the rational development and utilization of our existing land resources. And, to develop scientific and rational land use planning, it firstly needs an effective method of describing the land use change patterns and then to investigate the causes of these changes. All of the above can be attributed to land use analysis and optimization from the academic standpoint.

In the field of land use analysis and optimization study, Dr Liu Biao, Dr Huang Bo and Dr Zhang Wenting have done much research work, and made many significant achievements. They have now decided to publish a book on land use analysis and optimization. I have had the honor of reading the book and consider that it makes the following contributions to the theoretical research and practical application on land use analysis and optimization.

Firstly, three spatio-temporal logit models for land use change analysis, namely, the geographically and temporally weighted logit model (GTWLM), the spatio-temporal panel logit model (ST-PLM), and the generalized spatio-temporal logit model (GSTLM), are proposed accordingly to deal with the aforementioned issues. The GTWLM, which considers spatio-temporal non-stationarity, includes the temporal data in a spatio-temporal framework by proposing a spatio-temporal distance. The ST-PLM incorporates the spatio-temporal correlation and individual effect in one model, where the spatio-temporal correlation is considered in the random individual effect with an assumption that the correlation between such components is inversely proportional to the spatio-temporal distance. By integrating the GTWLM and the ST-PLM, the GSTLM explores the spatio-temporal non-stationarity and correlations simultaneously, whilst considering their individual effects in constructing an integrated model.

Secondly, a MOO-based two-level spatial planning of land use is proposed. Considering the spatial inconsistency between the municipal overall land use plan and the urban master plan in China, a MOO-based two-level spatial planning of land use is conducted. The spatial planning aims to manage and coordinate the land use to different geographic extents and involves spatial layouts and structures of land use at different levels. In spatial planning, the geographic information system (GIS) and remote sensing (RS) are used to evaluate, analyze, and measure environmental, economic, and social issues with regard to the spatial land use change. The quantitative relationships between these objectives and spatial land use allocation are then used as rules in the MOO process to simulate environmental conditions under different spatial land use allocation scenarios.

Achievement is founded on diligence and wasted by recklessness. Fortunately, there are many researchers who are working tirelessly on land use change study, such as the authors of this book. Because of them, we can finally achieve a scientific and rational development and utilization of land resources in order to solve the global problems of land resources and environmental protection. The publication of this book not only delivers their research achievements, but also the inherent requirements of social responsibility. I hope that more research institutions and scholars will make innovations and climb the scientific peaks bravely, so as to contribute to the field of land use change research.

Dr. Qiu Baoxing

Part 1

Introduction

Chapter 1

Introduction

Due to the drastic changes in land use, irrespective of their quantity or pattern, they have a profound influence on the people's daily lives. Unfortunately, because of the lack of scientific land use analysis and planning, there are bad influences which we do not accept. As a result, especially during the last half century, the importance of these two research fields has attracted the interests of researchers around the world. Although there have been numerous studies relating to them, this book analyzes and lists their several problems. Land use analysis and planning can often be studied respectively, but a combination of them is more and more closely sought.

So, in this introductory chapter, the detailed background of LUCC analysis and land use planning is introduced at the outset. Section 1.2 proposes several problems of the contemporary studies. And the objectives of this book are stated at the end of this chapter, in the hope of solving all the problems considered in the preceding sections.

1.1 Background

The 20th century has witnessed significant changes in land use patterns. During the last three centuries, almost 1.2 million km^2 of forest and woodland and 5.6 million km^2 of grassland and pasture have been converted to other uses, globally (Ramankutty & Foley, 1999; Alexander Popp *et al.*, 2016). Meanwhile, the impacts of land use change are escalating to threatening proportions around the globe. Many environmental problems can be traced back to land use change (Wolter *et al.*, 2015; Yao *et al.*, 2015): these include desertification, eutrophication, acidification, climate change, eustatic sea-level rise, greenhouse effect, and biodiversity loss. The study of land use change has, therefore, received a great deal of attention from policy-makers, planners, and developers who are seeking a sustainable land use plan (Turner & Meyer, 1991; Lambin *et al.*, 2003; Rindfuss *et al.*, 2004; Pontius *et al.*, 2007; Huang *et al.*, 2009a; Nicolas *et al.*, 2016). Both the International Geosphere-Biosphere Programme (IGBP) and the International Human Dimensions Programme (IHDP) have proposed a Land Use and Cover Change (LUCC) Core Project/Research Programme.

For the cities, one of the main factors of LUCC is urbanization. Urbanization is a common worldwide trend that is caused by population growth and economic development. In the urbanization process, social conflicts result in the mass movement of people, which undermine the operational systems of land access (Silberstein & Maser, 2000; Ngai *et al.*, 2016). Urbanization causes land degradation and uncontrolled natural resource use. China has witnessed astonishing economic growth and urban development during the past three decades. The rapid urban sprawl triggered by increased economic activity and population growth has caused numerous problems, such as the loss of open space, excessive carbon emission, severe soil erosion, and other environmental deterioration problems (Romero & Ordenes, 2004; Stone *et al.*, 2010; Ma *et al.*, 2016; Vanwalleghem *et al.*, 2017). Moreover, the competition of local, national, and international users with various socio-economic statuses and powers to achieve economic growth, and nature conservation, increases the difficulty of land use management. Therefore, both a scientific model to properly analyze land use change and a proactive plan to properly structure and allocate land use are required.

Land use change models aim to understand the causes and consequences of land use dynamics. Currently, several modeling approaches with different outcomes for land use change simulation and exploration are available. Such approaches offer possibilities for experiments to test our understanding of the key processes, thus allowing the model to describe the sensitivity of quantitative changes and provide alternative pathways into the future.

Earlier studies in land use modeling involve dividing the study area into grid cells and describing it by a pre-determined set of biophysical and socio-economic variables. Based on this, methods such as cellular automata (CA) (Clarke & Gaydos, 1998; Wu, 1998; Wu, 2002; Ahangaran, 2017), agent (Kohler & Gumerman, 2000; Gimblett, 2002), Markov chain analysis (Lopez *et al.*, 2001; Bielecki *et al.*, 2016), and logistic regression (Wu & Yeh, 1997; Cheng & Masser, 2003; Morotti & Grandi, 2016) are employed. The causal factors and the main models (CA, agent, regression, etc.) used for LUCC change analysis are described in the following sections.

Since land use change modeling endeavors to explore the relationship between causal factors and land use patterns or changes (multi-temporal land use patterns), the analysis of causal factors is an indispensable part in land use change modeling.

Land use change is a complex process influenced by a number of factors. As is evident from earlier studies, no single set of factors can explain the changes. Researchers (Turner, 1995; Bicik *et al.*, 2001) have provided a summarization of the causal factors. They have mentioned two kinds of factors: (1) natural factors, such as climate change, soil, hydrology and nature disasters, and (2) human factors, such as population, technology level, economic growth, and technology level. The IHDP report (Nunes *et al.*, 1999) also summarizes the causal factors into natural and human factors.

Initially, most researchers focused on natural factors, which were treated as the determinants to land use change. Crow *et al.* (1999) and Naveh (1995) considered the impact from climate change, soil, and DEM. Climate change was considered to be a major causal factor to land use change on a large spatial and temporal scale. The influence of natural factors on land use change is realized over a long time scale. However, human factors play a key role in land use change over a short time scale. Hence, more attention has been paid to the study of human factors. Many researchers consider economic growth as a very important factor (Bingham *et al.*, 1995; Houghton, 1994; Fischer *et al.*, 2001; Bicik *et al.*, 2001). Also the impact from population growth, political regimen, and technology development is studied (Reid *et al.*, 2000; Bicik *et al.*, 2001; Houghton, 1994). Nowadays, due consideration is given to both human and natural factors.

The important parameters that influence land use changes are as follows.

(1) demographic factors (population size, population growth, and population density) are widely treated as major causal factors of land use change (Verburg *et al.*, 2001);

(2) accessibility is also often viewed as a significant driver for land use change through its effect on the cost of transportation and ease of settlement (Geist & Lambin, 2001);

(3) spatial details play an important role in land use change processes (White *et al.*, 1997); and

(4) a causal force analysis conducted by Chen (2000) found that policy and economy generally influence land use change.

After examining relevant citations, Xie *et al.* (2006) provide a summary of causal factors commonly used in different land use change models, as follows:

Table 1.1 Causal factors for land use change

Category	Causal factor
Demography	Population size
	Population growth
	Population density
Proximity	Distance to road
	Distance to town/market
	Distance to settlement
	Distance to shopping mall
	Proximity to the urban structure
Economic	Investment structure
	Industry structure
	Housing commercialization
	Returns to land use (costs and prices)
	Job growth
	Cost of land use change
	Rent
Social	Affluence
	Human attitudes and values
Collective rule-making	Zoning
	Tenure
Site characteristics	Soil quality
	Slope
Constraints	Water body
	Environment-sensitive area
Neighborhoods	Availability of exploitable sites
	Agglomeration of developed areas
Others	Technology level

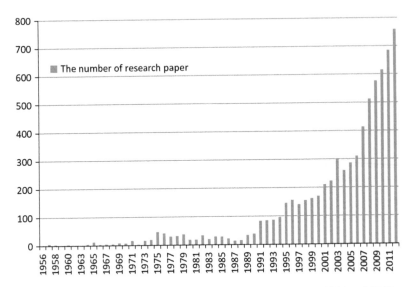

Figure 1.1 Increasing number of annual published research papers between 1956 and 2012
Topic: [land use or land-use] and [planning or plan]; Database: Science Citation Index Expanded (SCI-EXPANDED); Social Sciences Citation Index (SSCI); Years: 1956 to 2012.

Land use planning has become a common tool for managing land use resources and it has been widely studied (Hao, 2015). Figure 1.1 shows the number of published research papers on the topic of ("land use" or "land-use") and ("planning" or "plan") from 1956 to 2012. Since the 1960s, land use planning has had a long and commendable history of guiding the development of towns, cities, and regions (Berke, 2002). The number of published research papers significantly increased between 1991 and 2012, particularly after 2007, implying that there is an increasing significance of academic research concerning land use planning.

Figure 1.2 shows the spatial distribution of the cumulative number of published research papers between 1956 and 2012. Developed countries, such as the United States and Australia, as well as some European countries, provide a high proportion of the total number of published research papers. The results suggest that the degree of development influences the studies on land use planning.

Despite the significant number of studies on land use planning, some inconsistencies are still observed. Inconsistencies in the definitions of land use planning are most prominent. For instance, Leung (2003) proposed that land use planning is "a conception about the spatial arrangement of land use with a set of proposed actions to make that a reality" (p. 1). Stewart *et al.* (2004) defined land use planning as the process of allocating different activities or uses to specific areas within a region. According to The Canadian Institute of Planners (2000), land use planning is the scientific, aesthetic, and orderly disposition of land, resources, facilities, and services to secure the physical, economic, and social efficiency, as well as the health and well-being, of urban and rural communities (Bendor *et al.*, 2016). Several scholars have defined land use planning from the aspect of land use spatial allocation. Stewart *et al.* (2004) defined land use planning as "the process of allocating different activities or uses, such as industrial,

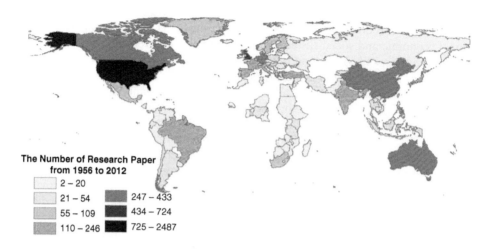

Figure 1.2 Total number of published research papers between 1956 and 2012 worldwide
Topic: [land use or land-use] and [planning or plan]; Database: Science Citation Index Expanded (SCI-EXPANDED); Social Sciences Citation Index (SSCI); Years: 1956 to 2012.

residential, commercial or recreation al activities, to spatial units of area within a region" (p. 1). Moreover, some scholars posit that land use planning offers a type of plan that orders and regulates the use of land. The AUMA (2007) indicated that land use planning is a key municipal function, which includes long-range land use policy, growth management, capital budgeting, and regulatory or implementation planning.

The variations in the definition are associated with the different goals required by various land use groups and the land use characteristics of a given study area. Evidently, uncertainty in definition likewise results in a complex and conflicting land use objective and constraint system. Based on the definitions proposed in existing academic studies, land use planning can be defined as the term used for a branch of public policy encompassing various disciplines on land use which seeks to achieve the land use goals and satisfy the land use constraints from different land use groups, thus preventing land use conflicts and ensuring benefit maximization across the entire land use system.

In China, various land use plans are used at different levels, and these land use plans are specifically ruled by government policies. The municipal overall land use plan and the urban master plan are two important land use plans. The municipal overall land use plan is one plan in the overall land use plan system that manages land use resources in an administrative scope at city level. The overall land use plan system comprises a series of plans from national level to town level. The municipal overall land use plan and the urban master plan have several overlapping functions because both plans are required in a given administrative scope based on Chinese policy. The urban master plan is concerned with the development of urban areas covered in the administrative scope but it disregards issues outside of the urban areas. For instance, the urban master plan may omit forest land use, which is inside the administrative scope but outside the urban area. Conversely, the municipal overall land use plan manages land use resources for the entire administrative scope, particularly agricultural protection and construction land growth control. The functions and scopes of these two plans in China are carefully regulated by policies.

The overall land use plan is an overall arrangement of land development, utilization, management, protection, and layout in space and time for a certain region, based on the national economic and social sustainable development requirements, as well as local, natural, economic, and social conditions. The overall land use plan is the foundation of national land use control in a certain region. The overall land use plan is the general arrangement for land use at the administrative and regional levels based on the characteristics of land resources and the socio-economic development requirements of subsequent periods (The State Council of the People's Republic of China, 2008). The overall land use plan is a highly comprehensive scientific work which must be conducted on the basis of reality. With the development of social economy and science and technology, the overall land use plan should be continuously revised and supplemented. The overall land use plan in China is divided into four levels, namely, the provincial level, the municipal level, the country level, and the town level. In this book we discuss the municipal overall land use plan.

The urban master plan guides the development of urban areas. The functions of the urban master plan involve environmental protection, economic development, and social equality within the scope of urban areas in an administrative scope. The scope of the municipal overall land use plan involves the spatial layout of land use in a region, the protection of the environment, and the protection of agricultural land and construction land growth.

In order to address the method/model of land use analysis and optimization in this book, the Chinese city of Shenzhen city is selected as the case study. In China, land use change is mostly noticeable in some of the big cities. Shenzhen is one of the most developed cities in China and the said city has grown at a dramatic speed. In less than 30 years, Shenzhen, formerly a tiny border town of 30,000 people in 1979, has developed into a modern metropolis. The land use pattern of Shenzhen has changed drastically in the past three decades. A careful study of the land use change in Shenzhen will benefit the development of other cities in China. Shenzhen's Special Economic Zone (SEZ) is an ideal choice for this study because of its "one city, two systems" framework.

1.2 Problem Statement

The main content of this book contains two parts; one is the land use change analysis/modeling, and the other is the land use planning. In this part, we mainly discuss the existing problems of these two research fields.

1.2.1 LUCC Analysis

As for land use change analysis, even though earlier land use change studies have demonstrated varying levels of success in their specific domains, several problems still need to be addressed.

First of all, with the limitation of land resource, not all the land use change is from rural to urban; more and more land use changes are occurring within the urban area. The modeling methodology should be capable of accommodating multi-class land use change (e.g., residential to commercial and industrial, etc. in support of urban redevelopment studies) rather than binary changes only (e.g., rural to urban in support of urban sprawl studies).

Secondly, due to different administrative, political, or other contextual issues, the relationships between land use change (or multi-temporal land use pattern) and causal factors in different locations and time may be intrinsically different. Hence, spatio-temporal non-stationarity, which indicates the different relationships which exist at different locations and different times, need to be considered in land use change analysis.

Thirdly, substantial amounts of spatial variables and spatial land use data tend to be self-dependent. This phenomenon is referred to as spatial autocorrelation (Legendre & Fortin, 1989). When dealing with spatio-temporal data, specific approaches should be considered to incorporate both spatial and temporal autocorrelation. Otherwise, inefficient parameter estimates and inaccurate measures of statistical significance will result.

Fourthly, each land cell is characterized by its own individual effect (*e.g.*, if the land type of a cell is urban, the probability of land use change in this cell will be lower than rural). Under such circumstances, a more detailed analysis of the individual effect in modeling land use change should be incorporated.

Finally, all the problems mentioned above are not totally disparate (*e.g.*, spatio-temporal non-stationarity can be partly considered in spatio-temporal autocorrelation). Thus, the land use change model should be capable of generalizing a series of changes in the spatio-temporal framework into a unified model for the better projection of land use distribution. This should be done with due consideration of multi-class land use type, spatio-temporal non-stationarity, spatio-temporal autocorrelations, and individual effect.

1.2.2 Land Use Planning

As for land use planning, there are three daunting tasks in land use planning in China: (1) the multi-objectives proposed by different land use groups; (2) the conflicts between the municipal overall land use plan and the urban master plan; and (3) the difficulty of implementing land use planning into space.

(1) Multi-objectives proposed by different land use groups

Land use planning provides rules and guidelines for land use management and the future development of a city. However, in reality, many objectives and goals of land use planning tend to conflict because of different emphases on the biophysical, ecological, or socio-economic aspects (Blaschke *et al.*, 1991; DeRose & North 1995; Guido, 2016). In land use systems, various groups (local, national, and international users with different socio-economic status and power) in society emphasize different constellations of these elements, depending on their own perspectives. Land use planning has become a multi-objective problem because of the various requirements from different land use groups, such as the government, businessmen, residents, and drivers. These proposed objectives are typically competing and conflicting. For instance, the objective of accommodating an increasingly urbanized population clashes with the objective of providing sufficient open space, which, in turn, clashes with the objective of ensuring a good level of transportation service. Moreover, multiple stakeholders have their own different priorities and solutions; therefore, rather than one best optimal solution, a set of trade-off solutions is required (Stewart *et al.*, 2004; Sorkhabi, 2016).

The increasing complexity of the land use planning problem is not only associated with the involvement of multiple objectives but also with the definition, measurement, and evaluation of the involved objectives (Yang *et al.*, 2014). These objectives may be unstructured, non-linear, and spatially related; thus, these objectives are difficult to handle. Within this context, computer-based techniques are required to assist planners in their decision-making.

Finally, considering that multiple objectives are major concerns in land planning systems compared with single objectives, trade-offs must be made on the basis of the relative importance of one objective against another (Li & Parrott, 2016; Biłozor *et al.*, 2014).

The process becomes more difficult because different groups have their own expectations on weighting factors for objectives (Balling *et al.*, 1999).

(2) Conflicts between the municipal overall land use plan and the urban master plan in China

Numerous planning systems, such as regional plans, plans for national economic and social development, water resource plans, and environmental plans, coexist in China (Lu *et al.*, 2004). All of these plans are created by different branches on the basis of various standards, and each plan has its own contents and functions. In reality, the interaction between these plans is unavoidable and causes problems (Datta & Regis, 2016), thus increasing the complexity of land use planning in China.

Among all of these plans in China, the municipal overall land use plan and the urban master plan are the most analogous because both plans oversee the spatial layout of land use in a city. The municipal overall land use plan conducts land use allocation in an administrative scope in a given city, whereas the urban master plan emphasizes the development of urban areas within an administrative scope. For instance, the municipal overall land use plan uses basic farmland protection as a major principle and attempts to control built-up land growth on the basis of the demand of land use resources from economic development, population growth, and environmental preservation. By contrast, the urban master plan is not involved in non-urban land use. The urban master plan concerns urban land use allocation within the urban area. The urban master plan aims to create a livable, accessible, and equitable urban living environment for the residents, and to provide sufficient housing and employment for the citizens.

In addition to differences in functions, the land use classification systems used by these two plans are different as well. The municipal overall land use plan usually covers eight land use types, namely, cultivated land, garden plot, forest, grassland, built-up land, transportation, water body, and unused land, whereas the urban master plan uses a more detailed land classification system in which built-up lands are classified as numerous types of urban land use.

Given the distinguishable differences, the connections between the municipal overall land use plan and the urban master plan are notable. The area of built-up land determined by the municipal overall land use plan is the geographic extent of the urban master plan in the future. The spatial extent is determined by the municipal overall land use plan based on a set of objectives and constraints. The urban master plan should be consistent with the results of the municipal overall land use plan and it should be conducted within the area of built-up land determined by the municipal overall land use plan. The built-up land extent determined by the municipal overall land use plan can influence the urban master plan because the total employment capacity, housing capacity, green space, and total available land that can be used for urban development are related to this extent. Therefore, these two plans are associated with each other to some extent. However, considering that these two types of plans are created by different branches, coordinating these two plans can be difficult, particularly in space.

(3) Difficulty of implementing land use planning into space

Managing the spatial layout of land use is one of the most important functions of all types of land use planning. However, in reality, achieving the land-use-related economic, social, and environmental indexes proposed by plans by spatial land use layouts is difficult. Therefore, even if plans have established objectives or guidance on managing land use and land use resources, meeting the goals of such plans by properly allocating land use in space is impossible. Spatially related objectives are impossible to achieve artificially because these objectives are typically non-linear and unstructured. Thus, computer-based techniques have been extensively used in land use planning systems to assist in the implementation of land use plans. However, a large gap still exists between setting goals and the reality of spatial land use allocation.

Land use planning in China is a significant issue, because the spatial extents of the municipal overall land use plan and the urban master plan partly overlap. The overlapped extent should be theoretically consistent. However, these two extents are usually inconsistent in reality. Spatial inconformity in China causes conflicts between the urban master plan and the municipal overall land use plan.

Within this context, the spatial planning of land use is required to coordinate the plans in space. The goal of spatial planning focuses on the spatial layout of land use and aims to manage land use plans at different levels, to attain the proposed objectives, and to satisfy the constraints in the urban master plan and the municipal overall land use plan.

1.3 Research Objectives

According to the key contents of this book, as in Section 1.2, addressing research objectives are also divided into two parts, LUCC analysis and land use planning, respectively.

1.3.1 LUCC Analysis Research

This book aims to develop a set of innovative models within the logistic regression framework to address the challenges identified and then to apply them in analyzing the land use changes in SEZ, Shenzhen. Specifically, the objectives are to:

(1)   extend the logit model to support multi-class land use change analysis, which implies a more complicated spatio-temporal effect;
(2)   extend the logit model to include spatio-temporal non-stationarity in multi-class land use change;
(3)   extend the logit model to include spatio-temporal autocorrelation in multi-class land use change;
(4)   extend the logit model to incorporate the individual effect for each land cell;
(5)   extend the logit model to generate a unified model for a time series of land use distributions, whilst considering spatio-temporal non-stationarity,

spatio-temporal autocorrelations, and the individual effect in multi-class land use change; and

(6) evaluate the performance and the reliability of the proposed approaches by using the land use data in SEZ, Shenzhen.

1.3.2 Land Use Planning Research

To address the problems of land use planning in China, this research aims to develop a two-level spatial planning of land use to coordinate the municipal overall land use plan and the urban master plan in space and to achieve the proposed land use planning objectives by using the MOO approach. In this study, the GIS and RS techniques are used to assist in evaluating and measuring the spatial objectives in planning. Given the multiple objectives, a MOO approach is used to create the trade-off between all of the objectives. Objectives that have been defined in spatial planning are proposed on the basis of local and national policies in the case study area and land use principles. Meanwhile, the research objectives of land use planning of this book are listed as follows:

(1) to define the land use planning objective system for spatial planning at the municipal overall land use plan level and the urban master plan level in accordance with different land use planning principles and local and national policies.

To achieve this objective, first, the land use principles are reviewed; secondly, the background of Shenzhen is subsequently provided, including the history, location, population, economy, land use changes, and the municipal overall land use plan and the urban master plan in Shenzhen; and, finally, some national policies, such as the Regulations on Basic Farmland Protection (The State Council, 1999), are used as references. Then, referring to the literature review on land use principles, the characteristics of Shenzhen, and some national or local policies, the objective system for spatial planning in Shenzhen will be defined;

(2) to measure and evaluate the complex spatially related objectives to ensure a better understanding of the effect of spatial land use changes, and then to evaluate these objectives in the spatial planning of land use.

After defining the objective system, all of these objectives will be measured and evaluated. Different spatial layouts and structures of land use have different effects on all of the objectives. The relationships between land use changes and objectives are established, in which the GIS and RS techniques have been used to map and analyze the spatial changes. Then, the MOO will calculate the fitness for each candidate plan based on the established relationships;

(3) to design an effective MOO model to balance land use conflicts.

The MOO has been well-developed for a long time. However, in this book, a two-level MOO, which connects the municipal overall land use plan and the urban master plan, has been designed. At the first level of this two-level MOO, the spatial layout and total amount of built-up land are provided based on multiple objectives. Then, based on the spatial layout of built-up land provided by the first level, at the second level, the optimization can be implemented within the geographical extent determined in the first level of optimization. In the MOO, the GA is used to search for the Pareto solutions, and a Maximin function is applied to calculate the fitness of the candidate plan;

(4)   to identify a set of uniform, recognizable, and comprehensive two-level spatial planning of land use, and then Pareto solutions are provided for planners and decision-makers.

The Pareto plans are provided by the designed approach. At the municipal overall land use plan level, spatial layouts and structures of land use are provided. Then, the planners can subsequently refer to the objective value of each Pareto solution and select one or several that they prefer. Given that the spatial layout at the municipal overall land use plan level has been determined, the built-up land at the municipal overall land use plan level will be considered as the geographical extent wherein the urban master plan will be implemented.

Chapter 2

Study Area

This chapter focuses on the brief introduction to Shenzhen as the case study area, which is the most important SEZ in China. Firstly, the basic situation of Shenzhen, as covered in this chapter, includes the history, location, population, economic development, land use changes, and environmental problems. Secondly, the land use change in Shenzhen is analyzed, and the results evince that planning plays an important role in the urbanization of Shenzhen. Finally, causal factors for land use change analysis are proposed.

2.1 Basic Situation

2.1.1 History and Location

Shenzhen (22° 27'N to 22° 52'N, 113° 46' E to 114° 37'E) is located in the eastern part of the Pearl River Delta Region and it is the only land crossing into Hong Kong (Figure 2.1), with a total terrestrial area of 1,952.84 km². Considering its proximity to Hong Kong, Shenzhen was developed as an SEZ in 1979 (Bruton *et al.*, 2005). Since then, it has experienced rapid urbanization and achieved significant economic development (Chen *et al.*, 2012). From a small fishing town, Shenzhen is now one of China's major economic centers(Chen *et al.*, 2012), and it is on the list of top global cities (Lv *et al.*, 2011).

Shenzhen has ten districts, namely, Futian, Luohu, Baoan, Longgang, Yantian, Guangming, Longhua, Longgang, Pingshan, and Dapeng (see Figure 2.2). Prior to 2010, Luohu, Nanshan, Yantian and Futian were the original SEZs located near to Hong Kong. In 2010, the State Council of the People's Republic of China approved the inclusion of Baoan and Longgan as SEZs (Shi & Yu, 2014). That same year, Baoan included the new district of Guangming, and Longgang included the new district of Pingshan (see Table 2.1).

2.1.2 Population and Economy

Designed as the first SEZ in China, Shenzhen grew from a city with a population of approximately 0.3 million in 1980 to one with 8.9 million inhabitants in 2009. Statistical data confirm that Shenzhen experienced rapid population growth. Some practising planners have even declared that Shenzhen's population has already reached

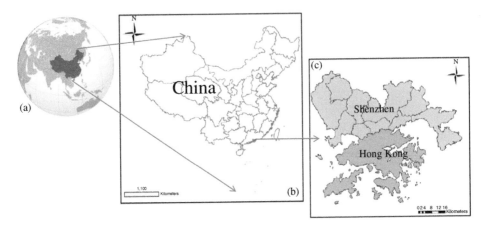

Figure 2.1 Research area (a) location of Shenzhen in China on Earth, (b) location of Shenzhen in China, and (c) administrative districts map of Shenzhen and the location of Hong Kong

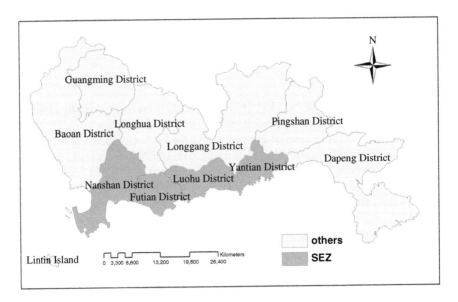

Figure 2.2 Distribution of districts and the SEZ in Shenzhen

14 million (Ng, 2011), a figure relatively larger than the one provided in the statistical data.

Meanwhile, with a GDP of 196 million RMB in 1979, Shenzhen has attained an annual growth rate of 25.8% for the past 30 years. In 2009, the GDP of Shenzhen reached 820 billion RMB (Ng, 2011). Figure 2.3 shows the GDP development and population growth along with the employed population of Shenzhen from 1980 to 2008.

Table 2.1 Administrative division of Shenzhen in 2010

Chinese Name*	English Name
福田区	Futian
罗湖区	Luohu
盐田区	Yantian
南山区	Nanshan
宝安区 (含光明新区)	Baoan (including Guangming)
宝安区 (不含光明新区)	Baoan (excluding Guangming)
光明新區	Guangming
龙岗区 (含坪山新区)	Longgang (including Pingshan)
龙岗区 (不含坪山新区)	Longgang (excluding Pingshan)
坪山新区	Pingshan

* The first column is the name of districts in Chinese.

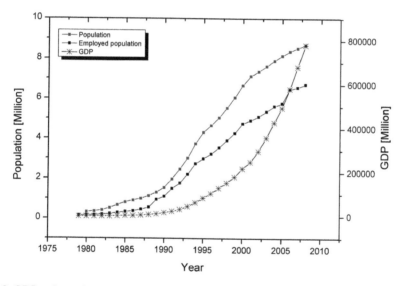

Figure 2.3 GDP and population growth of Shenzhen from 1980 to 2008
Source: Shenzhen Statistics Bureau (2009).

2.2 Land Use Change Information

As mentioned already, Shenzhen is the fastest-growing city in China, with its land use is experiencing dramatic change. The primary type of land use change is from non-built-up to built-up. As can be observed from Figures 2.4, 2.5, and 2.6, the built-up area in Shenzhen increased from 41,613.24 hectares to 92,050.32 hectares in 18 years. Most of this change has occurred in Baoan and Longgang (outside the SEZ). In these two districts, the built-up area has increased from 28,269.32 hectares to 71,383.28 hectares. The increase in the built-up areas in the SEZ is only 7,323.12 hectares. That is because most of land in the SEZ has already been used. As can be seen from Figure 2.4,

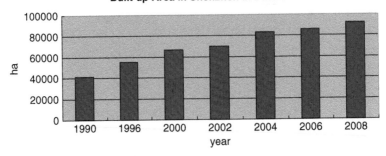

Figure 2.4 Built-up area in Shenzhen in different years
Source: Urban Planning, Land and Resources Commission of Shenzhen Municipality.

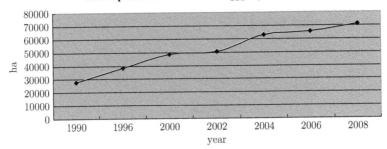

Figure 2.5 Built-up area in Baoan and Longgang in different years
Source: Urban Planning, Land and Resources Commission of Shenzhen Municipality.

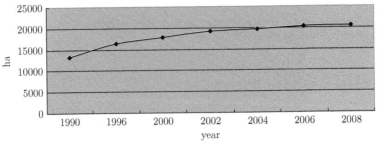

Figure 2.6 Built-up area in the SEZ in different years
Source: Urban Planning, Land and Resources Commission of Shenzhen Municipality.

the curve representing the increasing trend of built-up areas in the SEZ has become flattened. The land use change, which may happen in residential, industrial, commercial, and transportation, is more complex. This study is performed in the SEZ and it will take into account the multi-class land use change.

At the same time, land use maps for the years 1990, 1996, 2000, 2002, 2004, 2006, and 2008 were acquired from land investigation and digital orthophotos obtained from the Urban Planning, Land, and Resources Commission of Shenzhen Municipality. Each land cell is classified into one of the eleven land use types: farmland; woodland; garden land; grass land; residential; industrial; commercial; financial; education; transportation: and water land use type. Because the land uses of commercial, financial, and education are too small in their proportions, the eleven land use types were aggregated into four categories in order to focus on the human development of land. These four categories are: undeveloped; residential; industrial; and commercial/transportation/others land use type (excluding water). Figures 2.7(a) ~ (g) represent the land use maps corresponding to 1990, 1996, 2000, 2002, 2004, 2006, and 2008, respectively. All maps have the same size of 1,801 columns by 1,002 rows and a common spatial resolution of 50 m. All the maps follow the same legend: green is for undeveloped; yellow is for residential; purple is for industrial; red is for commercial/transportation/others; blue is for water; and white is for "no data". It can be found from Figure 2.7 that most of the industrial land use is concentrated in Nanshan whereas undeveloped land use types are prominent in Yantian. The residential and commercial/transportation/others land use are distributed randomly in Luohu, Futian, and Nanshan. Water bodies are treated as constraints that are inappropriate for development.

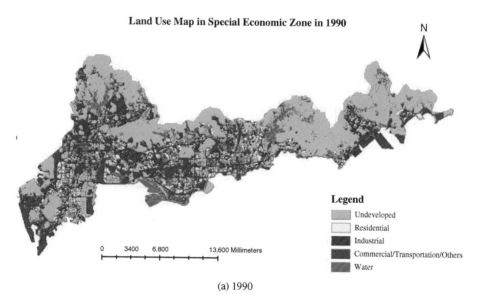

(a) 1990

Figure 2.7 SEZ land use maps
Source: Urban Planning, Land and Resources Commission of Shenzhen Municipality.

Land Use Map in Special Economic Zone in 1996

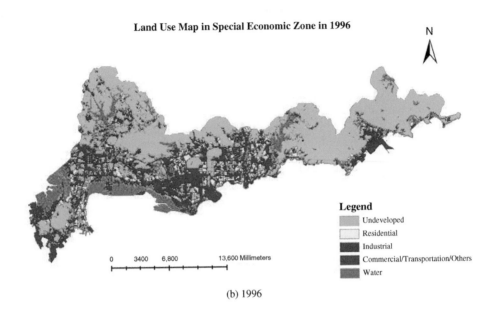

(b) 1996

Land Use Map in Special Economic Zone in 2000

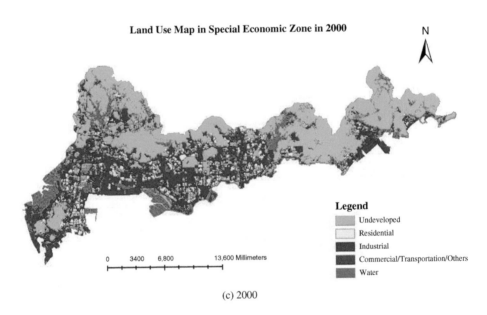

(c) 2000

Figure 2.7 (cont.)

Land Use Map in Special Economic Zone in 2002

N

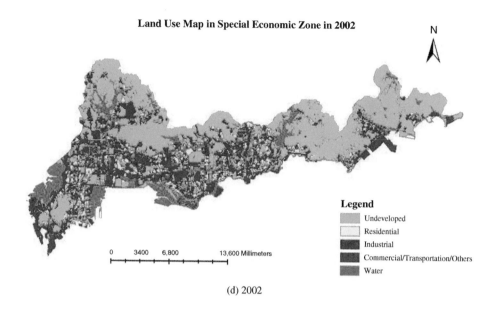

Legend

Undeveloped

Residential

Industrial

Commercial/Transportation/Others

Water

0   3400   6,800   13,600 Millimeters

(d) 2002

Land Use Map in Special Economic Zone in 2004

N

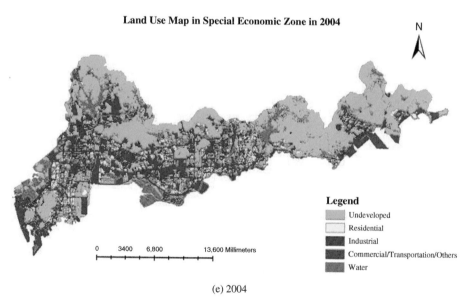

Legend

Undeveloped

Residential

Industrial

Commercial/Transportation/Others

Water

0   3400   6,800   13,600 Millimeters

(e) 2004

Figure 2.7 (cont.)

Land Use Map in Special Economic Zone in 2006

N

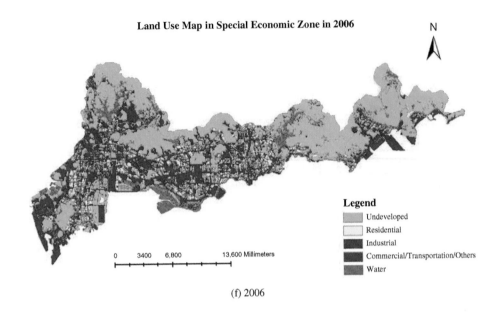

Legend

Undeveloped

Residential

Industrial

Commercial/Transportation/Others

Water

0   3400   6,800   13,600 Millimeters

(f) 2006

Land Use Map in Special Economic Zone in 2008

N

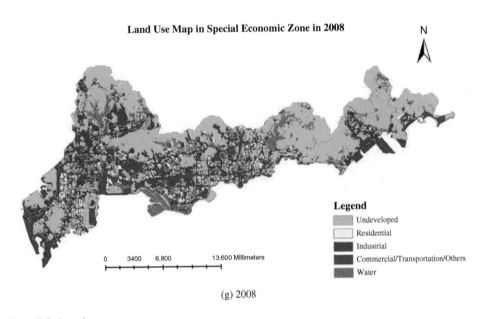

Legend

Undeveloped

Residential

Industrial

Commercial/Transportation/Others

Water

0   3400   6,800   13,600 Millimeters

(g) 2008

Figure 2.7 (cont.)

2.3 Causal Factors for Land Use Change Analysis

Considering previous literature, the context of SEZ, and the data availability, this study includes the following causal factors data in the proposed model: elevation data; transportation data (major roads and railroads); commercial map; financial map; industrial map; educational facilities map; demographic data; and planning map (1996–2010). All these shape files/raster layers were compiled in ESRI ArcMap v9.1[®]. All layers were projected to the *WGS_1984_W114* coordinate system. Raster layers were resampled by using a cell size of 50 meters to ensure consistency with other land use data.

Slope raster was generated from the elevation data by using the ArcMap spatial analyst extension. Sequential shape files for the road networks, rail lines/stations, commercial centers, financial centers, industrial centers, educational facilities were generated from land investigation data by feature selection. Subsequently, the ArcMap spatial analyst extension was used to generate distance raster layers to these utilities in each modeled year according to the Euclidian distance.

The demographic data corresponding to each district in Shenzhen was obtained from the statistical year book at the Shenzhen Library. In order to use the available demographic data, spatialization of the demographic data was done by Dr Li Hongga from Institute of Remote Sensing Applications, CAS. Then, spatial interpolation method called Kriging was employed to obtain the population density distribution.

The planning map (1996–2010) in the CAJ format was obtained from the Urban Planning, Land and Resources Commission of Shenzhen Municipality. As can be found from Figure 2.8, there are many layers in the CAJ map and the road network is

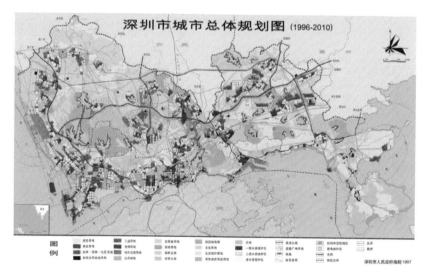

Figure 2.8 Shenzhen planning map
Source: Urban Planning, Land and Resources Commission of Shenzhen Municipality.

not closed. Thus, the CAJ map could not be converted to shapefile directly. The land use layer for each land use type planning should be extracted from the CAJ map and the road work should be closed. The land use planning is also classified into five types in accordance with the land use map. Then, the land use planning map and road network planning can be made. Both the maps are shown in Figures 2.9 and 2.10,

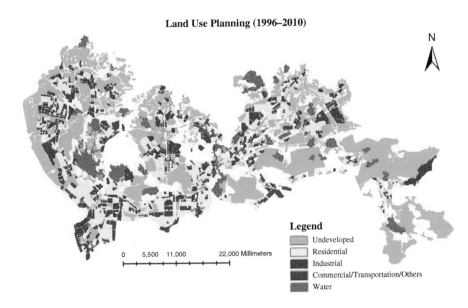

Figure 2.9 Land use planning in Shenzhen

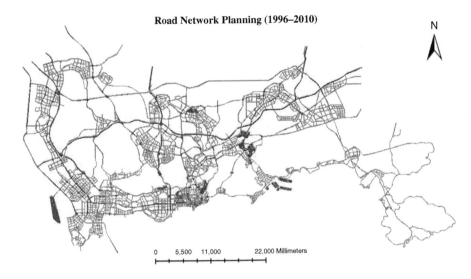

Figure 2.10 Road network planning in Shenzhen

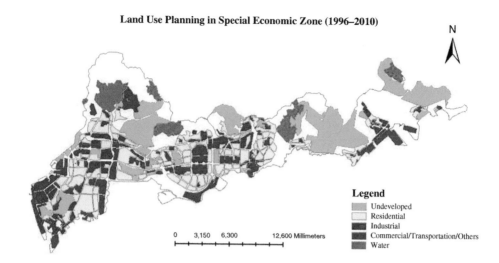

Figure 2.11 Land use planning in the SEZ

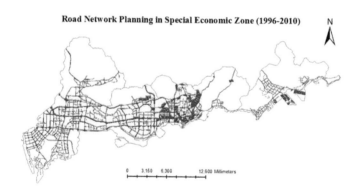

Figure 2.12 Road network planning in the SEZ

respectively. For our research, the two maps are clipped to the SEZ shape (Figures 2.11 and 1.12).

Overall, the framework of driving factors preparation is shown in Figure 2.13.

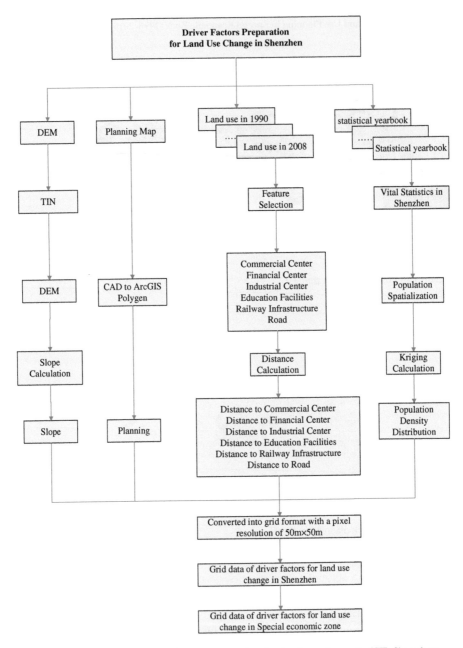

Figure 2.13 The framework of driver factor preparation for land use change in SEZ, Shenzhen

Table 2.2 Summary of explanatory variables for the land use change model

Variables	Definition
Distance to commercial center	Distance from the cell to the commercial center
Distance to financial center	Distance from the cell to the financial center
Distance to industrial center	Distance from the cell to the industrial center
Distance to educational facilities	Distance from the cell to the nearest educational facilities
Distance to railway infrastructure	Distance from the cell to the railway infrastructure
Distance to road	Distance from the cell to the nearest road
Population	Population density of the cell
DEM	DEM
Slope	Measurement of the degree of slope
Planning	Land use planning

Overall, ten causal factors were considered in this study. A summary of these factors is shown in Table 2.2.

Chapter 3

LUCC Analysis and Optimization Model Review

In this chapter, a variety of techniques for land use change modeling are briefly reviewed. Especially, the spatio-temporal model of land use change is introduced. Although many land use change models have been developed, some researchers also conclude that none of the approaches yet "dominates this nascent field". Owing to the complexity of the land development process and the differences in modeling objectives, an obviously superior approach is hard to find. Each method has its strengths, weaknesses, and particular application domains. With due consideration to the pros and cons of the abovementioned models, the logistic regression framework has been selected in this book. Despite the fundamental assumptions problem, the regression model is still considered to be a powerful tool for land use change analysis.

As for optimization studies, the chapter firstly reviews many optimization approaches for the spatial planning of land use. LP is a traditional technique, which was proposed in the 1960s, but it has not been especially suitable for land use planning issues. Such issues are generally multi-objective optimization problems. Their objectives are usually paradoxical, and lead to complex computing. Many studies have proved that heuristic algorithms, which have shown good performance in some respects, are good at solving MOO problems. In this chapter, we focus on genetic algorithms (GA) which have been applied to land use planning. And, subsequently, to evaluate the goodness of a plan, the fitness evaluation method is presented. It is good at selecting excellent ones from many solutions in a complex MOO problem.

3.1 Land Use Change Analysis Models

3.1.1 Cellular Automata

Tobler (1979) proposed using cellular automata (CA) as a tool for modeling spatial dynamics. Couclelis (1985, 1988, 1997) explored the implications of the idea in a series of theoretical papers. The approach has been implemented by others in a wide range of applications (*e.g.*, Batty & Xie, 1994; Benenson, 1998; Ceccini & Viola, 1990; Clark *et al.*, 1997; Papini & Rabino, 1997; Phipps, 1989; Portugali & Benenson, 1995; White & Engelen, 1993, 1999; White, Engelen, & Uljee, 1997). Recently, the approach has been linked to GIS and applied to land use change study.

CA are dynamic spatial systems in which the state of each cell in an array depends on the previous state of the cells within the cell's neighborhood, based on a set of state transition rules. As the system is discrete and iterative and involves interactions only within local regions rather than between all pairs of cells, it is therefore possible to work with grids containing substantial cells. The very fine spatial resolution that can be attained is an important advantage when modeling land use dynamics. Conventional CA each consist of:

(1)   a Euclidian space divided into an array of identical cells;
(2)   a neighborhood of defined cell size and shape;
(3)   a set of discrete cell states;
(4)   a set of transition rules, which determine the state of a cell as a function of the states of cells in the neighborhood; and
(5)   discrete time steps, wherein all cell are states updated simultaneously.

However, these defining characteristics can be interpreted broadly, or relaxed in response to the requirements of a particular modeling problem. Hence, many types of CA can exist.

CA offer a number of advantages. As already mentioned, they permit extreme spatial detail. Hence, they are able to reproduce the actual complexity which occurs in nature. Besides, owing to the high resolution and raster nature, they are compatible with GIS databases, and can be linked with them in a relatively simple manner. At the other end of the spatial scale, CA can be linked through their transition rules to other, macro-scale models that limit or drive the CA dynamics. This facilitates the comprehensive modeling of integrated environmental–human systems. Finally, CA are defined and calibrated in a single operation, since calibration corresponds to finding the optimal transition rules. This means that CA models are typically implemented much faster and simply than traditional spatial models.

CA also have some disadvantages. Many CA models inductively assume that the human impact is important, but they do not explicitly model decisions. Others explicitly hypothesize a set of agents coincident with lattice cells and they use transition rules as proxies to decision-making. These efforts succeed when the unit of analysis is tessellated, decision-making strategies are fixed, and heterogeneous actors are affected by local neighbors in a simple, well-defined manner. However, when actors are not tied to a location in the intrinsic manner of CA cells, the problem of spatial orientedness might arise (Hogeweg, 1988). "Spatial orientedness" refers to the extent to which neighborhood relationships do not reflect actual spatial relationships. This can be remedied by using techniques that have non-uniform transition rules and can dynamically change the strength and configuration of the connections between the cells. Because these characteristics are beyond the capacities of the rigidly defined CA, the pure, traditional CA method may not be generally suited to model land use/cover change. Furthermore, CA models focus on the simulation of spatial patterns rather than on the interpretation of the spatio-temporal land use change process.

3.1.2 Multi-agent System

While CA models focus on transitions, agent-based models focus on human actions. Agents are the basic component in these models. Several characteristics of these agents are described as follows:

(1)   agents are autonomous;
(2)   agents share an environment through agent communication and interaction; and
(3)   agents make decisions that link behavior to the environment.

Agents are used to represent numerous entities, including animals, people, and organizations. Agents must act in accordance with various rules, which link their autonomous goals to the environment. An autonomous agent needs to enable it react to environmental changes. Beyond pure reaction, some of the models are based on rational choice theory. These models generally assume that the actors are perfectly rational optimizers. Consequently, these agents are able to maximize their well-being and can balance long-run *vs* short-run payoffs.

A large number of scholars attempt to employ agent-based models in land use/cover change study. In the context of a land use change model, an agent may represent a land manager who can decide a land use. The model agents may also represent entities or social organizations such as a village assembly, local governments, or a neighboring country. Agent-based models in land use/cover change study consist of the decision rules, such as income maximization, of each human actor, and their environmental feedbacks. Land markets, social networks, and resource management institutions may provide other significant environments for interaction. When coupled with a cellular model representing the landscape on which the agents act, these models are truly compatible for explicitly representing the land use change processes.

However, the appropriateness of models of perfect rationality for agent-based modeling applied to land-use/cover change remains questionable, given the importance of spatial interdependencies and feedbacks in these systems. Recognition of the complex environment in which human decision-making occurs has resulted in a movement towards agent-based models that employ some variant of bounded rationality (Gigerenzer & Todd, 1999). Typically, bounded rational agents have goals that relate their actions to the environment. Rather than implementing an optimal solution that fully anticipates all future states of the system they constitute, they make inductive, discrete, and evolving choices that move them towards achieving goals (Bower & Bunn, 2000).

3.1.3 Markov Chain Analysis

In mathematical terminology, a Markov chain is a discrete random process. A discrete random process refers to a system that exists in various states and changes randomly in discrete steps. Because the system changes randomly, it is generally impossible to predict the exact future state of such a system. However, the statistical properties of the system in the future can often be described.

In land use change, different land use categories represent the states of a chain. The Markov chain states that the probability of land use category at time t, X_t, (and in fact during all future steps) depends only on the land use category at time t -1, X_{t-1}, and not additionally on the land use category of the system during the previous steps. The changes in the state of land use are called transitions, and the probabilities associated with various state-changes are called transition probabilities. The probability of a land use change from a land use category (state) a_i to a land use category (state) a_j in one time period, "one step transition probability", is $P(X_t = a_j | X_{t-1} = a_i)$. In the case of a homogeneous Markov chain, the transition probability can be expressed as:

$$P(X_t = a_j | X_{t-1} = a_i) = P_{ij} \tag{3.1}$$

The transition probability can be estimated by using the following equation:

$$P_{ij} = \frac{n_{ij}}{n_i} \tag{3.2}$$

where n_{ij} is the number of times the land use changed from state i to j, and n_i is the number of times that land use category a_i occurred.

Combining all the transition probabilities between all states $a_1, a_2, \ldots, a_m$ gives the following transition matrix:

$$\mathbf{P} = (P_{ij}) = \begin{bmatrix} P_{11} & P_{12} & \cdots & P_{1m} \\ P_{21} & P_{22} & \cdots & P_{2m} \\ \vdots & \vdots & & \vdots \\ P_{m1} & P_{m2} & \cdots & P_{mm} \end{bmatrix} \tag{3.3}$$

According to the Chapman–Kolomogorov equation, the n steps transition matrix can be easily deduced from the one-step transition matrix:

$$\mathbf{P}^{(n)} = \mathbf{P}^n \tag{3.4}$$

Therefore, the land use state of a period after n steps can be easily predicted.

In Markov chain analysis, the transition probabilities are estimated as the proportions of cells that have changed state from one point in time to another. It is a useful way of estimating these probabilities despite the development of procedures for estimating transition probabilities on the basis of more complex scientific consideration. However, the Markov chain model lacks explanatory power; this is because the causal relationships underlying the transition studies have been left unexplored.

3.1.4 Expert Models

Expert models combine expert judgment with non-frequentist probability techniques such as Bayesian probability, the Dempster-Schaefer theory (Eastman, 1999), or symbolic artificial intelligence approaches such as expert systems and rule-based knowledge systems (Gordon & Shortliffe, 1984; Lee *et al.*, 1992). These methods express qualitative knowledge in a quantitative fashion that enables the modeler to determine where specified land uses are likely to occur. However, it can be difficult to incorporate all aspects of the problem domain, and this leaves room for gaps and inconsistencies.

3.1.5 Evolutionary Models

Within the field of artificial intelligence, symbolic approaches such as expert systems are complemented by using the biologically inspired evolutionary paradigm. Artificial neural networks and evolutionary programming have already found their way into LUCC models (*e.g.*, Balling *et al.*, 1999; Mann & Benwell, 1996). In brief, neural networks are silicon analogs of neural structures that are trained to associate outcomes with stimuli. Evolutionary programming mimics the techniques of Darwinian evolution by employing computational algorithms/programs that iterate over many generations to generate the solutions for a problem.

3.1.6 Logistic Regression

In statistics, regression analysis is the collective term for techniques used for modeling and analyzing numerical data consisting of values of a dependent variable and of one or more independent variables. In the aforementioned process, the dependent variable is also called the response variable or measurement and the independent variable is called the explanatory variable or predictor. The dependent variable in the regression equation is modeled as a function of the independent variables, corresponding parameters ("constants"), and an error term. The error term is treated as a random variable. It represents the unsubstantiated variation in the dependent variable. The parameters are estimated to give a "best fit" of the data. Typically, the best fit is evaluated by using the least squares method, but other criteria have also been used. For land use change analysis, the logistic regression model is widely applied to estimate the outcomes of a categorical dependent variable, whereas the independent variables can be a combination of both continuous and categorical variables. It is a suitable approach with which to estimate the coefficients of casual factors from the observation of land use change; this is because the determinants of land use change usually consist of both continuous and categorical variables.

Considering urbanization, for instance, a collection of m independent variables will be denoted by the vector $X' = (x_1, x_2, \ldots, x_m)$. Let the conditional probability that the conversion of rural lands to urban and other built-up uses is happening be denoted by $P(y = 1|X)$, or simply by $P(X)$. The linear logistic model is represented as follows:

$$\text{logit}[P(X)] = \ln\left(\frac{P(X)}{1 - P(X)}\right) = \alpha + \beta'X + \varepsilon \qquad (3.5)$$

where α is the intercept, $\beta = (\beta_1, \beta_2, \ldots, \beta_m)'$ is the corresponding factor coefficient vector, and ε is the independent error term, that is, $E(\varepsilon_i, \varepsilon_j) = 0$.

To fit the logistic regression model in Equation 3.5 to a set of data requires that the value of α and β_is, the unknown parameters, are estimated. Unlike ordinary linear regression, the method of ordinary least squares (OLS) estimation cannot be applied to a model with a dichotomous outcome. Instead, maximum likelihood techniques are used to maximize the value of a function, the log-likelihood function. The work by Hosmer and Lemeshow (1989) provides more detailed information. Finding the solution of the likelihood equations entails special software, which may be found in several packaged programs. Different combinations of explanatory variables are available for the regression model, such as stepwise regression, "best subset" models, and predefined conceptual models.

Earlier studies considered the spatial dependence of the land use data when employing logistic regression to model land use change analysis. This can generally be done by building a model that incorporates an autoregressive structure (Anselin, 1988). For example, a spatial lag model can be used as an extension of the traditional linear regression model:

$$y = a + \sum_{i=1}^{m} x_i b_i + \rho W y + \varepsilon \qquad (3.6)$$

where ρ is a coefficient on the spatially lagged dependent variable and W is a spatial weight matrix. It should be noted that the maximum likelihood estimator (MLE) is usually employed to solve the parameters of such a model that best fits the data:

$$L = \left[y \ln\left(\frac{\exp(a + X\beta + \rho W y)}{1 + \exp(a + X\beta + \rho W y)}\right) - (1 - y)\ln(1 + \exp(a + X\beta + \rho W y)) \right] \qquad (3.7)$$

The likelihood can be maximized by using a simplex unvaried optimization routine (LeSage, 1998). The logistic version of Model, which incorporates spatial autocorrelation, called the Spatial Autologistic Regression model has been devised (Dubin, 1995; Dubin, 1997; LeSage, 1998). This model, also called the Spatial AutoLogit (SAL) model, has been proven to be effective in regression involving spatial autocorrelation (Páez & Suzuki, 2001). Similar to the spatial lag model, a spatial error model (Anselin, 1988) can also be utilized for regression models involving spatial autocorrelation. Nonetheless, they are primarily used for diagnostic analysis rather than for extrapolation-like prediction (Jetz et al., 2005).

Logistic regression improves gradually and allows the causal factors to be a mixture of continuous and categorical variables, which suits land use change data well. However, the assumption of logistic regression, such as the uncorrelation of the error part, is not always valid. If the assumption cannot be satisfied, the generalization performance of logistic regression may degrade drastically.

3.2 LUCC Optimization

3.2.1 Optimization Method

Optimization approaches were used in land use planning. The linear programming (LP) was a traditional optimization technique type where both the objectives and the constraints were linear and additive (Greenberg, 1978). The LP model was first articulated in the 1960s to solve the linear or quadratic equations that addressed the problems in the urban planning systems (Aerts et al., 2003; Kaur et al., 2004). Kaur et al. (2004) integrated the LP with the hydrologic model. They further proposed a land use optimization plan that considered the land use effects on conservation. Ligmann-Zielinska et al. (2008) developed a novel multi-objective spatial optimization model based on LP. This model encouraged efficient urban land use utilization by infill development, compatibility of adjacent land uses, and defensible redevelopment. These mathematical programming methods were generally straightforward and powerful. However, these methods often required extensive additional efforts in recasting the problem in a feasible framework. Generally, the methods were not applicable for large-scale planning applications because of the intolerable amount of computing time and cost. The LP model cannot handle non-linear and unstructured objectives such as spatial problems in land use. Furthermore, multiple objectives were involved in the land use planning system. Hence, LP was efficient only when a single objective was clearly identified (Stewart *et al.*, 2004). The LP technique was not any more suif for land use planning issues.

Within this context, multi-objective optimization approaches were proposed to address MOO problems in land use or urban systems from as early as the 1970s. Bammi and Bammi (1979) employed the classic weighted sum method to minimize conflict between adjacent land use, travel time, tax cost, and other objectives. They created a land use plan for DuPage County in the USA by using the MOO approach. Barber (1976) and Arad and Berechman (1978) discussed the land use design in terms of the units of different activities that were allocated to each regional zone. They also considered interactions between regions. Barber reported an application of one of the earliest interactive algorithms for solving MOO problems. Laterand Hopkins (1977) described an assignment model of land use planning. This model included a quadratic assignment term that represents the interactions between land uses in different sites. However, none of these earlier MOO studies of land use planning dealt with space-related environmental objectives, such as non-point source pollution, and soil erosion.

More studies have recently attempted to develop MOO approaches that measure spatial attributes and deal with various kinds of spatial information. Ligmann-Zielinska *et al.* (2008) developed a novel multi-objective spatial optimization model based on LP. This model encourages efficient urban space utilization through infill development, compatibility of adjacent land uses, and defensible redevelopment. Chandramouli (2007) used a similar approach for land use allocation optimization but added a visualization tool to the MOO to assess the Pareto optimal plans. Masoomi *et al.* (2013) used the multi-objective particle swarm optimization algorithm to find the parcel-level optimum distribution of urban land use. The algorithm simultaneously considered multiple objectives and constraints. The integration of GIS into most of these studies was a commonly popular choice to measure and present the spatial issues in MOO. Cao *et al.* (2011) used the multi-objective land use model coupled with a revised version of the genetic where the GIS was used to represent the diversified land use scenarios.

Aside from the spatial problems that have received increased attention, developing adaptive approaches in the area of MOO has attracted growing interest. GAs and other heuristic algorithms have received a great deal of attention as alternatives in recent decades. The GAs generated various alternative plans by using crossover and mutation operations. These algorithms developed a trade-off set of non-dominated plans. The GA was a robust and efficient general global optimization algorithm used to search for large, complex, and little-understood search spaces such as those of the multi-objective land use planning problems (Zhang *et al.*, 2010). The GA worked with a population of plans. Instead of offering one "best" solution, a number of Pareto optimal solutions were generated. This set of alternative solutions was well suited to practical applications and provided options for planners to choose from. Another alternative land use plan was selected from the pool of Pareto optimal solutions if implementing an optimal was difficult or impossible (Srivastava *et al.*, 2002). Many researchers applied GA to solve multi-objective land use planning problems. Accordingly, some meaningful outcomes were achieved. Balling *et al.* (1999) used a GA with a maximum fitness function to generate a Pareto optimal set of future land use and corridor-upgrade plans. They concluded their work after the Pareto set was derived. Selection of the final solution from the Pareto plans still needed further investigation. Stewart *et al.* (2004) used a GA to incorporate multiple objectives in land use planning. They further discovered that the heuristic algorithm invalidated the use of traditional LP models as an optimization approach for land use planning systems. Brookes (2001) used a GA connected with a region-growing program that generated alternative patch configurations. The GA effectively resolved multiple patch problems, thereby making this algorithm better than the LP method.

The MOO approaches of land use were widely studied. Many meaningful outcomes were achieved. The optimization techniques were developed from the LP to the heuristic approach (*e.g.*, GA). The land use system objectives were developed from a single, linear, and non-spatial objective to multiple, non-linear, and spatial objectives. Moreover, the Pareto solutions were proposed to widen the decision-maker's choices. Finally, the integration with GIS and RS made spatial objective evaluation easy. The results of the MOO are evident. Within the Chinese context, however, some issues in the MOO application for spatial land use planning still need further consideration.

3.2.2 Fitness Evaluation

Fitness is an indicator by which to evaluate the goodness of a plan in one generation. Therefore, the higher the fitness value, the better the plan (Balling, 2003; Balling *et al.*, 2004; Lowry & Balling, 2009). The fitness helps us to select the father and mother plans. And the father and mother plans with high fitness are expected to generate the children with high fitness. Therefore, by the iterations of GA, the fitness of the whole generation becomes larger, and this means that the plans have been optimized. In order to achieve the optimization by GA, we need to devise a fitness function to measure the fitness of each plan.

There are numerous means to compute fitness, like ranking, normalized sum objectives, and weighted average of normalized objectives (Hajela & Lin, 1992; Konak *et al.*, 2006). In this study, the Maximin fitness function proposed by Balling (2002) is employed to measure the goodness of each plan in one generation. Firstly, translate all objectives into the format of "min(Z)", and then let Ob_{ki} as the value of the k-th objective in the i-th plan. As for the max(Z) format objective, the objective will be transformed min(Z) format by the following equation.

$$Z = -Z \tag{3.8}$$

Now consider two plans in one generation, the i-th plan and the j-th plan. The i-th plan will be dominated by the j-th plan if:

$$Ob_{1i} > Ob_{1j}, Ob_{2i} > Ob_{2j}, \ldots, Ob_{ki} > Ob_{kj} \tag{3.9}$$

And this equation is equivalent to the following equation:

$$min(Ob_{1i} - Ob_{1j}, Ob_{2i} - Ob_{2j}, \ldots, Ob_{ki} - Ob_{kj}) > 0 \tag{3.10}$$

Thus, the i-th plan is a dominated plan if:

$$\max_{i \neq j} \left(min(Ob_{1i} - Ob_{1j}, Ob_{2i} - Ob_{2j}, \ldots, Ob_{ki} - Ob_{kj}) \right) > 0 \tag{3.11}$$

And the fitness of the i-th plan is:

$$f_i = \left[1 - \max_{j \neq i} \left(\min\left(\frac{Ob_{1i} - Ob_{1j}}{Ob_{1-max} - Ob_{1-min}}, \ldots, \frac{Ob_{ki} - Ob_{kj}}{Ob_{k-max} - Ob_{k-min}}\right)\right)\right]^p \quad (3.12)$$

In above equations, the scaling factors Ob_{k-max} and Ob_{k-min} are the maximum and minimum value of the k-th objective. According to Equation 3.12, the fitness of Pareto-optimal plans will be between one and 2^p, whereas the fitness of dominated plans will be between 0 and 1. As for the exponent p, if it is larger than 1, it will make the fitness of the Pareto-optimal plans even higher and the fitness of the dominated plans even lower. In Balling's study, he used a high value of p which was 15, and it made the GA quite aggressive in pursuing Pareto-optimal solutions (Balling *et al.*, 1999).

Part II

Spatio-temporal Analysis Model

Chapter 4

Classic MNLM Model

In this chapter, the formation of the MNLM is covered. MNLM has a good performance on data in which the dependent variable is discrete and unordered, and in which the independent variables are continuous or categorical predictors. Subsequently, PCP and Kappa coefficients, which are used to evaluate the classification accuracy and the ability of the model, are discussed. It measures the overall concordance between a classification and the actual land use type, and it is employed to assess the goodness-of-fit of the models. Finally, the land use change in the SEZ is analyzed and the results evince that planning plays an important role in the urbanization of the SEZ.

4.1 Multinomial Logit Model (MNLM)

An MNLM is used for data in which the dependent variable is discrete and unordered, and in which the independent variables are continuous or categorical predictors. Unlike a binary logistic model, wherein a dependent variable has only a binary choice (*e.g.*, persist/change), the dependent variable in an MNLM can have more than two choices that are coded categorically, and one of the categories is assumed to be the reference category.

Assuming that y_i, the dependent variable for the ith observation, has $J + 1$ categories and is indexed as j ($j = 0,1, \ldots, J$). If y_i *is* j, let $y_i^j = 1$, otherwise $y_i^j = 0$. Let x_i denote the exploratory variables for the i-th observation. Category 0 is considered to be the reference category. The logits, which compare any category $j = 1, \ldots, J$ with the reference category, are described as follows:

$$\log\left(\frac{prob(y_i^j = 1)}{prob(y_i^0 = 1)}\right) = \log\left(\frac{\exp(\beta^j x_i)}{\exp(\beta^0 x_i)}\right) = \beta^j x_i, \quad j = 1, \ldots, J \tag{4.1}$$

where β^j is the coefficient vector of length p, corresponding to p covariates. β^0 is set to be zero. $\beta^1, \ldots, \beta^J$ are to be estimated. The parameter, β^j, represents the additive effect of a one-unit increase in the independent variable, x, on the log-odds of being in category j, rather than in the reference category. An alternative way of interpreting the effect of an independent variable, x_i, is to use predicted probabilities $prob(y_i^j = 1)$ for different values of x_i:

$$prob(y_i^j = 1) = \frac{\exp(\beta^j x_i)}{\sum_{j=0}^{J} \exp(\beta^j x_i)}, j = 1, \ldots, J \tag{4.2}$$

Then, the probability of being in the reference category, "0" (*normal*), can be calculated by subtraction:

$$prob(y_i^0 = 1) = 1 - \sum_{j=1}^{J} prob(y_i^j = 1) \tag{4.3}$$

In this model, the same independent variable appears in each of the j categories, and $\beta^1, \ldots, \beta^J$ is usually estimated for each contrast.

The maximum likelihood method was employed to estimate the MNLM. Firstly, the probabilities were calculated by Equations 4.2 and 4.3 for each cell's category and time each other. Then, the parameters can be obtained by maximizing the product.

4.2 Evaluation of the Model's Classification Accuracy

4.2.1 Percentage of Correct Prediction (PCP)

The percentage of correct prediction (PCP), which measures the overall concordance between a classification and the actual land use type, is employed to assess the goodness-of-fit of the models. An efficient way to assess the goodness-of-fit of classification is utilized in this study, which cross-tabulates predictions with observations and calculates the overall concordance. Table 4.1 provides an example of the cross-evaluation table used in this study. 0, 1, 2, and 3 represent undeveloped, residential, industrial, and commercial/transportation/others land use type, respectively.

Let us consider land use type 0 and 1 as an example. $Num(0-0)$ is the number of cells with a label of 0 (undeveloped) and is classified as 0 in Table 4.1. $Num(0-1)$ is the number of cells with a label of 0 (undeveloped) but is classified as 1 (residential). $Num(1-0)$ is the number of cells with a label of 1 (residential) but is classified as 0. $Num(1-1)$ is the number of cells with a label of 1 (residential) and is classified as 1.

$Num(P-0)$, $Num(P-1)$, $Num(P-2)$, and $Num(P-3)$ are the number of cells with a label of 0, 1, 2, and 3, respectively. $Num(O-0)$, $Num(O-1)$, $Num(O-2)$

Table 4.1 An example of the cross-evaluation table

Observed	Predicted				Total
	0	1	2	3	
0	Num(0–0)	Num(0–1)	Num(0–2)	Num(0–3)	Num(O–0)
1	Num(1–0)	Num(1–1)	Num(1–2)	Num(1–3)	Num(O–1)
2	Num(2–0)	Num(2–1)	Num(2–2)	Num(2–3)	Num(O–2)
3	Num(3–0)	Num(3–1)	Num(3–2)	Num(3–3)	Num(O–3)
Total	Num(P–0)	Num(P–1)	Num(P–2)	Num(P–3)	Num(Total)

and $Num(O-3)$ are the number of cells that are classified as 0, 1, 2, and 3, respectively. $Num(Total)$ is the size of the training set/evaluation set. Based on the cross-evaluation table, some important indicators can be calculated. For example:

$$PCP = \frac{Num(0-0) + Num(1-1) + Num(2-2) + Num(3-3)}{Num(Total)} \quad (4.4)$$

$$PCPo0 = \frac{Num(0-0)}{Num(O-0)} \quad (4.5)$$

$$PCPp0 = \frac{Num(0-0)}{Num(P-0)} \quad (4.6)$$

PCP indicates the overall accuracy of the classifier. $PCPo0$ is the probability that the land use category 0 can be accurately predicted by the model. It reveals the capacity of the classifier to detect the land use categorized as 0. The greater the value of $PCPo0$, the more the land use 0 can be correctly predicted. $PCPp0$ is measured by the percentage of correctly predicted land use 0 over the total land cells classified as 0. It exhibits the efficiency of the classifier in detecting the land use category 0. A higher $PCPp0$ implies that the model can predict the land use 0 with a higher accuracy.

4.2.2 Kappa

Kappa, which is usually attributed to Cohen (1960), is a member of a family of indices to measure the agreement between cellular maps. The indices (Cohen's Kappa) are given by Equation 4.7:

$$Kappa = \frac{PCP - p_c}{1 - p_c} \quad (4.7)$$

where PCP is the observed proportion agreement between the raters, and p_c is the expected proportion agreement due to chance. Kappa has the following properties: if the observed correct proportion is greater than the expected correct proportion due to chance, then Kappa > 0; if the observed correct proportion is equal to the expected correct proportion due to chance, then Kappa = 0; and if the observed correct proportion is less than the expected correct proportion due to chance, then Kappa < 0. If the raters are in complete agreement, then Kappa = 1.

The Cohen's Kappa is appropriate for contingency tables when the scientist does not have control over the marginal distributions. In contrast, usually the goal of a spatially explicit model is to obtain similar marginal distributions. Thus, Pontius and Gilmore (2000) suggest the use of Kno to evaluate a model's overall success, Klocation to evaluate the model's ability to specify location, and Kquantity to evaluate the model's ability to specify quantity. Suppose that there are J categories, Table 4.2 give the proportion of correct classification according to a model's ability to specify quantity and location accurately.

Table 4.2 Proportion correct classification according to a model's ability to specify accurately the quantity and location

Ability to Specify Quantity	Ability to Specify Location		
	No(NL)	Medium(ML)	Perfect(PL)
No(NQ)	$1/J$	$(1/J) + \text{Klocation}(NQPL-(1/J))$	$\sum_{j=0}^{J-1} \min(1/J, O_j)$
Medium(MQ)	$\sum_{j=0}^{J-1}(P_j, O_j)$	Percentage of Correct Prediction	$\sum_{j=0}^{J-1}(P_j, O_j)$
Perfect(PQ)	$\sum_{j=0}^{J-1}(O_j^2)$	PQNL + Klocation(1-PQNL)	1

where $O_j = \frac{Num(O-j)}{Num(Total)}$ $j = 0, 1, 2, 3$ and $P_j = \frac{Num(P-j)}{Num(Total)}$ $j = 0, 1, 2, 3.$

Following that, the definitions of Kappa coefficients are given as follows:

$$Kno = \frac{PCP - NQNL}{1 - NQNL} \tag{4.8}$$

$$Klocation = \frac{PCP - MQNL}{MQPL - MQNL} \tag{4.9}$$

$$Kquantity = \frac{PCP - NQML}{PQML - NQML} \tag{4.10}$$

The model's performance is corroborated from the values of the Kappa coefficients, which are closer to 1.

4.3 Results and Discussion

The regression result generated by MNLM, estimated by using the maximum likelihood algorithm, is shown in Table 3.4. In this study, the negative effect of the distance variable for residential land use reflects the fact that the residential zones are increasingly dispersed into the locations where the living environment is more attractive. Almost all the distance variables have a negative effect on the industrial land use type, except on the distance to the educational facilities and the distance to road. For transportation/commercial/others, the transportation factors (distance to the rail infrastructure and the distance to road) have significant negative effect. This reflects that transportation infrastructure plays a crucial role in the expansion of transportation/commercial/others.

The positive signs on population density for all built-up land use types suggest that population is a chief factor influencing urbanization. Meanwhile, the signs on DEM and slope are significantly negative, which implies that the probability of built-up land use type will decrease as the elevation and slope increase.

Industrial zoning regulations are not estimated to have any significant effect on any land use types. Residential and transportation/commercial/others zoning appear to have a positive impact on such development. Because the "no planning" factor is not significant for all land use types, such (no planning) places appear in a random development pattern. In all, zoning appears to mainly facilitate the development of corresponding land uses, but does not necessarily impede other land use types. Zoning policy continues to play a pivotal role in controlling the urban development in the SEZ.

Chapter 5

Geographically and Temporally Weighted Logit Model

Typically, non-stationarity is a characteristic of spatial data. Non-stationarity refers to the variations of the relationship between the response variable and the explanatory variables in a regression model. Fotheringham and Brunsdon *et al.* (1996) proposed geographically weighted regression (GWR) to consider spatial non-stationarity by estimating the parameters in each location. This chapter aims to explore the possibility of extending GWR to support a discrete model in a spatio-temporal setting.

In this chapter, non-stationarity is introduced. Following that, various ways in which earlier research has dealt with spatial or temporal non-stationarity are described and analyzed. Then, a basic framework for GWLM is presented and it is extended to include temporal data as GTWLM, which is applied to the study of land use change in SEZ, Shenzhen. Finally, the results for different models are compared and analyzed and it is concluded that the proposed model (GTWLM) performed better than the other models.

5.1 Introduction

Applied work in spatial analysis always relies heavily on the sample data that are collected with reference to location. When the sample data have a locational component, spatial heterogeneity may occur in the relationships which are being modeled. This could violate the basic assumption of statistical independence of observations, which typically is required for unbiased and efficient estimation (Huang *et al.*, 2009). A thorough understanding of non-stationarity is inevitable to utilize the literature for considering spatial and temporal non-stationarity in land use change modeling. The term "stationarity" is often taken to refer to the outcome of some process that has similar properties at all points of interest. In other words, the statistical properties (*e.g.*, mean and variance) of the variable or variables do not vary over the area of interest. Irrespective of the location, a stationary model has the same parameters, whereas with a non-stationary model the parameters are allowed to vary locally.

The traditional approach for land use change modeling is to estimate the underlying function globally. However, it is obvious that spatio-temporal non-stationarity exists in land use change models. Herein, acknowledgement of the stationary assumption is perhaps likely to be invalid; this is because parameters tend to vary over space and time. For example, the expansion of the European Union with the consequent likely restructuring of the Common Agricultural Policy (CAP) will lead to shifts in the political and economic forces driving LUCC in the region. The implication is that some factors, which are contributors to land use change models (*e.g.*, socio-economic factors), are likely to interpret land use change differently at varying locations and points of time.

Meanwhile, local models which consider the non-stationarity have been a source of immense interest. This is due to the better results such local models could provide than the global models. Numerous local models have been developed, such as the spatial expansion method, spatially adaptive filtering, multilevel modeling, Geographically Weighted Regression (GWR), dynamic model, and coefficient smoothing. However, due attention has not been paid to spatial and temporal non-stationarity within the same framework. Thus, based on the logistic regression model, this chapter aims to construct a geographically and temporally weighted logit model (GTWLM) including both spatial and temporal non-stationarity to assist the spatio-temporal land use change analysis. The emphasis on the logistic regression model is due to the fundamental role it plays in examining the relationship quantitatively between land use changes and explanatory variables. A study of spatio-temporal land use change in SEZ, Shenzhen will be carried out to evaluate the performance and reliability of the proposed model. A significant improvement achieved by GTWLR will be demonstrated by comparing the estimation results of the models.

5.2 Local Models

Interest in local forms of spatio-temporal analysis and spatio-temporal modeling is not new. The local models can be constructed in various ways. This section reviews several local models, which are capable of considering both spatial and temporal non-stationarity.

5.2.1 Models Considering Spatial Non-stationarity

5.2.1.1 The Spatial Expansion Method

The spatial expansion model which considers spatial non-stationarity was introduced by Casetti (1972):

$$y_i = \beta_1 x_{i1} + \ldots + \beta_p x_{ip} + \varepsilon_i \tag{5.1}$$

where y denotes a dependent variable, the x are independent variables, $\beta_1, \beta_2, \ldots, \beta_p$ represent parameters to be estimated, and ε represents an error term. This global model can be expanded by allowing each of the parameters to be functions of the coordinates. For instance, the parameters are allowed to vary geographically, as follows:

$$\beta_{1i} = \alpha_{10} + \alpha_{11} u_i + \alpha_{12} v_i \tag{5.2}$$

$$\beta_{2i} = \alpha_{20} + \alpha_{21} u_i + \alpha_{22} v_i \tag{5.3}$$

$$\beta_{pi} = \alpha_{p0} + \alpha_{p1} u_i + \alpha_{p2} v_i \tag{5.4}$$

where u_i and v_i represent the spatial coordinates of location i. Equations 5.2 to 5.4 represent simple linear expansions of the global parameters. This expansion method is very important in promoting the awareness of spatial non-stationarity. However, the form of the expansions equations needs to be assumed to follow *a priori*.

5.2.1.2 Spatially Adaptive Filtering

Another approach that allows parameters to vary is spatial adaptive filtering (Widrow & Hoff, 1960):

$$y_i = \beta_1 x_{i1} + ... + \beta_p x_{ip} + \varepsilon_i \tag{5.5}$$

As a rule, when a new observation occurs at another location j, the existing regression parameters, $\beta_1, \beta_2, ..., \beta_p$, are used to predict the dependent variable y_j. In spatially adaptive filtering, the values of the regression coefficient $\hat{\beta}$ are adjusted to $\hat{\beta}^j$ to improve the estimate. To avoid problems of overcompensation, the degree of adjustment applied could be "damped" by the following update rule:

$$\hat{\beta}^j = \hat{\beta} + |\hat{\beta}| \alpha \left(y_j - \hat{y}_j \right) / |\hat{y}_j| \tag{5.6}$$

where $\hat{y}_j$ is the predicted value of y_j based on $\hat{\beta}$ and α is the damping vector controlling the extent to which the correction is applied for parameters. In spatial analysis, the flow of coefficient estimates updating is a collaborative process between a pair of neighboring zones. This requires iterating between coefficient estimates until some form of convergence is achieved. Then, the results should be a unique estimate of the regression coefficient vector β for each case. The fact remains that the case-wise correction procedure is some degree of spatial smoothing of the estimates of individual elements of β. Thus, the method tends only to allow the parameters to "drift" slowly across the geographical space.

5.2.1.3 Multilevel Modeling

Multilevel modeling tries to combine an individual-level model with a macro-level model. The model has the following form:

$$y_{im} = \alpha_m + \beta_m x_{im} + \varepsilon_{im} \tag{5.7}$$

where y_{im} represents the value of individual i living place m, and x_{im} is the ith observation of attribute x at place m. α_m and β_m are rewritten as follows:

$$\alpha_m = \alpha + \mu_m^\alpha \tag{5.8}$$

$$\beta_m = \beta + \mu_m^\beta \tag{5.9}$$

where μ_m^α and μ_m^β are random place-specific variables. Substituting Equations 5.8 and 5.9 into Equation 4.7 yields the multi-level model,

$$y_{im} = \alpha + \beta x_{im} + \left(\varepsilon_{im} + \mu_m^\alpha + \mu_m^\beta x_{im} \right) \tag{5.10}$$

Adding place attributes μ_m^α and μ_m^β into α_m and β_m, and extending the number of levels in the hierarchy are the salient advantages of this model. However, the reliability of the *a priori* definition of a discrete set of spatial units at each level is a disadvantage.

5.2.1.4 Geographically Weighted Regression (GWR)

The GWR model allows regression parameters to be estimated locally. The model can be expressed as follows:

$$Y_i = \sum_p \beta_p(u_i, v_i)X_{ip} + \varepsilon_i \ i = 1, \ldots, n \tag{5.11}$$

where (u_i, v_i) denotes the coordinates of the point i in space, and $\beta_p(u_i, v_i)$ represents a set of parameters at point i.

The estimation of parameters $\beta_p(u_i, v_i)$ is given by the following equation:

$$\hat{\beta}(u_i, v_i) = (X^T W(u_i, v_i)X)^{-1} X^T W(u_i, v_i)Y \tag{5.12}$$

where $W(u_i, v_i)$ is a $n \times n$ diagonal matrix computed for each point i. The closer the observation is to i, the greater the weight:

$$W(u_i, v_i) = \begin{pmatrix} w_{i1} & 0 & \ldots & 0 \\ 0 & w_{i2} & \ldots & 0 \\ \ldots & \ldots & \ddots & \vdots \\ 0 & 0 & \ldots & w_{in} \end{pmatrix} \tag{5.13}$$

Especially, GWR based on the regression model has a powerful explanatory capability and it facilitates exploring the variation of the parameters by visualization. This implies that the local parameter estimates are capable of being mapped. However, the estimations required for each point increase the computational burden of this method.

5.2.2 Models Considering Temporal Non-stationarity

Models built during each time period can partly consider temporal non-stationarity. Besides, as can be seen below, there are some other methods which deal with this issue.

5.2.2.1 Dynamic Model

The dynamic model is used to articulate and model the behavior of the system over time. In such a model, the state in time t is usually derived from the state before time t. This framework facilitates the carefully consideration of temporal non-stationarity.

For example, in the article entitled "Modeling Land Use Decisions with Aggregate Data: Dynamic Land Use", a Markov decision process is used to represent the conditional land use decisions made by landowners. The decision process is assumed to be non-stationary so that the probability that the cell transforms from one land use to another changes over time. Also, it is dependent on the output and input prices as well as on other relevant decision variables. The dynamic modeling framework could be used to recover the non-stationary transition probabilities for land use allocations.

The principal advantage of using dynamic models is the ability to identify the land use change components in the future. Often, this information is needed to accurately represent the responses to and the impacts of the land use policies. Summarily, although estimating dynamic models is complex and time-consuming, such models have powerful capabilities for prediction.

5.2.2.2 Coefficient Smoothing

In this method, for each explanatory variable, the relationship between its influence coefficient and time is investigated by using a certain function. The function could be a quadratic polynomial or exponential smoothing and so on. Huang *et al.* (2009) used a model coupled with a logistic regression model and an exponential smoothing technique for exploring the effects of various factors on land use change. The modified exponential smoothing technique is employed to generate a smoothed model from a series of bi-temporal models obtained from different time periods.

By integrating the bi-temporal models, this method generalizes the details of land use change over different phases during the long period. The smoothing technique can be implemented at the coefficient level or the utility function level. The challenge here is that smoothing at the coefficient level entails a specific relationship, *e.g.*, linear, which may not be present.

5.3 Methodology

5.3.1 Overview of the Geographically Weighted Logit Model (GWLM)

Assuming that there is a set of n observations x_{ip} with the spatial coordinates (u_i, v_i), $i = 1, 2, \ldots, n$, on p predictor variables, $p = 1, 2, \ldots, P$, and a set of n observations on a dependent variable y_i^j, which is an indicator of the observed land use type j for cell i: if yes, y_i^j equals 1; if no, y_i^j equals 0. j is the index of land use type in this study. The underlying model for GWLM is as follows:

$$log\left(\frac{prob(y_i^j = 1)}{prob(y_i^0 = 1)}\right) = \sum_p \beta_p^j(u_i, v_i)X_{ip} + \varepsilon_i \tag{5.14}$$

where $j = 0$ denotes the undeveloped land use type as the basic land use type in this study, $\beta_p^j(u_i, v_i)$ are p continuous functions of the location (u_i, v_i) in the study area. Unlike the MNLM, the GWLM allows the parameters to vary across space, and hence it is more likely to capture positional effects.

Parameter estimation is a moving window process. A region or window is drawn around a location i, and all the data points within this region or window are used to estimate the parameters. The process is repeated for each observation in the data, and, finally, a set of parameter estimates is obtained for each location. The estimator of $\beta_p^j(u_i, v_i)$ is given at each location i by using the multinomial logistic regression with X's transform, as follows:

$$X' = W(u_i, v_i)^{1/2}X \tag{5.15}$$

where $W(u_i, v_i)$ is an n by n diagonal matrix:

$$W(u_i, v_i) = \begin{pmatrix} w_{i1} & 0 & \cdots & 0 \\ 0 & w_{i2} & \cdots & 0 \\ & & \ddots & \vdots \\ \cdots & \cdots & & \\ 0 & 0 & \cdots & w_{in} \end{pmatrix} \tag{5.16}$$

In the global weighted regression models, the values of $W(u_i, v_i)$ remain constant. However, it is assumed that the observed data close to point i have a greater influence in $W(u_i, v_i)$ than the data located farther from the point of observation i in the GWLM.

Basically, there are two kernels to construct weighting regimes: the fixed kernel, and the adaptive kernel. For the fixed and the adaptive kernel, the distance and the number of neighbors remain constant, respectively. Frequently used weighting functions include the bi-square function, the tri-cube kernel function, the Gaussian function, and the exponential distance decay based function. The exponential distance decay-based function, which is employed in this case, is expressed as follows:

$$w_{im} = \exp\left(- \frac{d_{im}^2}{h^2}\right) \tag{5.17}$$

where h is the bandwidth, and d_{im}, which is the spatial distance between location i and m, can be calculated by Equation 5.18:

$$d_{im} = \sqrt{(u_i - u_m)^2 + (v_i - v_m)^2} \tag{5.18}$$

where (u_i, v_i) and (u_m, v_m) refer to the point coordinates. According to Equations 5.17 and 5.18, the weighting of the other data decreases in accordance with the exponential form when the distance between i and m increases. If i and m coincide, the weighting value at this point will be equal to 1.

There is an issue involved in the estimation of GWLM. On the one hand, compared to OLS, multinomial logistic regression, which involves non-linear optimization, is time-consuming. Employing the entire sample will greatly increase the computational load. Hence, when parameters are estimated in every point, sub-samples are chosen by using a computational ruse that ignores observations with negligible weight. On the other hand, a sub-sample cannot be produced with the values of all single types (e.g., all ones). Hence, the weighting method employed here should guarantee the estimation of multinomial logistic regression. Adaptive weighting functions are used to adapt themselves in size to ensure that the same and enough number of non-zero weights are used for each multinomial logistic regression point being analyzed. In this manner, the kernels have larger bandwidths where the data points are sparsely distributed and have smaller ones where the data are abundant. In this study, the specific adaptive weighting function is expressed as follows:

$$w_{im} = \begin{cases} \exp\left(- \frac{d_{im}^2}{h^2}\right), & \text{if } m \text{ is the } q \text{ nearest points around } i \\ 0, & \text{otherwise} \end{cases} \tag{5.19}$$

where h stands for the bandwidth, q represents the proportion of observations to consider in the estimation of regression at each location.

The choice of appropriate bandwidth h and sub-sample size q is very important for the GWR-based model. Then, the weighting matrix could be decided by cross-validation. Actually, it can be automatically obtained with an optimization technique by maximizing PCP (Hurvich et al., 1998; Fotheringham et al., 2002).

5.3.2 Geographically and Temporally Weighted Logit Model (GTWLM)

The form of the GTWLM is as follows:

$$log\left(\frac{prob(y_i^j = 1)}{prob(y_i^0 = 1)}\right) = \sum_p \beta_p^j(u_i, v_i, t_i)X_{ip} + \varepsilon_i \tag{5.20}$$

where t_i is temporal coordinate, and $\beta_p^j(u_i, v_i, t_i)$ are p continuous functions of the spatio-temporal coordinates (u_i, v_i, t_i) in the study area.

In order to estimate $\beta_p^j(u_i, v_i, t_i)$ in each point, X should be transferred into X' in GTWLM, as follows.

$$X' = W^{ST}(u_i, v_i, t_i)^{1/2}X \tag{5.21}$$

$W^{ST}(u_i, v_i, t_i)$ is also a diagonal matrix, whose elements w_{im}^{ST} should be the spatio-temporal distance decay functions of (u, v, t) when calibrating the weight between point m adjacent to observation point i. Consequently, the estimation of GTWLM relies on the appropriate specification of the spatio-temporal distance decay function. Nevertheless, space and time are usually measured in different units and have different scale effects. To handle the scale problem, all the coordinate data are firstly standardized. For the difference about scale effects, an ellipsoidal coordinate system could be employed to combine the spatial effect and the temporal effect into the same framework. This is done to construct the spatio-temporal distance between the point, which will be estimated, and its surrounding observed points (Figure 5.1).

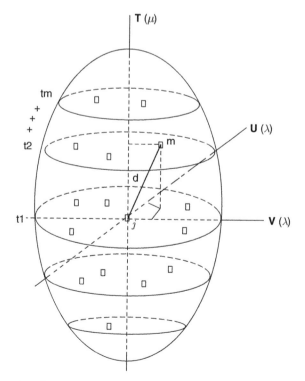

Figure 5.1 An illustration of spatio-temporal distance

Where i is regression point, m is the nearby point (u_m, v_m, t_m), and $m = \{1, 2, \ldots, n\}$ $(m \neq i)$. The formula of the spatio-temporal distance is as follows:

$$d_{im}^{ST} = \sqrt{\lambda((u_i - u_m)^2 + (v_i - v_m)^2) + \mu(t_i - t_m)^2} \tag{5.22}$$

where t_i and t_m are the temporal coordinates, and λ and μ are the scale factors to balance the different effects used to measure the spatial and temporal distance in their respective metric systems. The d_{im}^{ST} will be adopted to compute the spatio-temporal distance in GTWLR.

Based on d^{ST}, the elements of $W^{ST}(u_i, v_i, t_i)$, w_{im}^{ST} is defined.

$$w_{im}^{ST} = \begin{cases} \exp\left\{ -\left(\dfrac{(d_{im}^{ST})^2}{h_{ST}^2} \right) \right\}, & \text{if } m \text{ is the } q \text{ nearest points around } i \\ 0, & \text{otherwise} \end{cases} \tag{5.23}$$

where h_{ST} is the spatio-temporal bandwidth. Specifically, we substitute Equation 5.22 in 5.23:

$$w_{im}^{ST} = \exp\left\{ -\left(\frac{\lambda[(u_i - u_m)^2 + (v_i - v_m)^2] + \mu(t_i - t_m)^2}{h_{ST}^2} \right) \right\}$$

$$= \exp\left\{ -\left(\frac{(u_i - u_m)^2 + (v_i - v_m)^2}{h_S^2} + \frac{(t_i - t_m)^2}{h_T^2} \right) \right\}$$

$$= \exp\left\{ -\left((d_{im}^S)^2 \Big/ h_S^2 + (d_{im}^T)^2 \Big/ h_T^2 \right) \right\}$$

$$= \exp\left\{ -(d_{im}^S)^2 \Big/ h_S^2 \right\} \times \exp\left\{ -(d_{im}^T)^2 \Big/ h_S^T \right\}$$

$$= w_{im}^S \cdot w_{im}^T$$

where $w_{im}^S = \exp\left\{ -(d_{im}^S)^2 \Big/ h_S^2 \right\}$, $w_{im}^T = \exp\left\{ -(d_{im}^T)^2 \Big/ h_S^T \right\}$, $(d_{im}^S)^2 = (u_i - u_m)^2 + (v_i - v_m)^2$, $(d_{im}^T)^2 = (t_i - t_m)^2$, $h_S^2 = \frac{h_{ST}^2}{\lambda}$ and $h_T^2 = \frac{h_{ST}^2}{\mu}$ are the spatial and temporal bandwidths, respectively. Since the weighting function is a diagonal matrix, whose diagonal elements are multiplied by $w_{im}^S \cdot w_{im}^T$ $(1 \leq m \leq n)$, W^{ST} can be seen as the combination of spatially weighted matrix W^S and temporally weighted matrix W^T : $W^{ST} = W^S \times W^T$. Thus, if μ is set to 0, GTWLM becomes GWLM, which considers only the spatial non-stationarity. If λ is set to 0, GTWLM becomes the temporal weighted logit model (TWLM), which considers only the temporal non-stationarity.

Similar to GWLM, the parameters are also estimated by the CV method. In order to reduce the parameters to be estimated in the model, Equation 5.22 is divided by $\sqrt{\lambda}$ $(\lambda \neq 0)$.

$$d_{im}^{ST} / \sqrt{\lambda} = \sqrt{[(u_i - u_m)^2 + (v_i - v_m)^2] + \tau(t_i - t_m)^2} \tag{5.25}$$

where τ represents the parameter ratio μ/λ and can be seen as a balance ratio between the spatial distance effect and the temporal distance effect. Thus,

$$
\begin{aligned}
w_{im}^{ST} &= \exp\left\{-\left(\frac{\lambda[(u_i - u_m)^2 + (v_i - v_m)^2] + \mu(t_i - t_m)^2}{h_{ST}^2}\right)\right\} \\
&= \exp\left\{-\left(\frac{[(u_i - u_m)^2 + (v_i - v_m)^2] + \tau(t_i - t_m)^2}{h_{ST}^2/\lambda}\right)\right\}
\end{aligned}
\tag{5.26}
$$

Here, setting $\lambda = 1$ will not change the results estimated by λ, μ, and h_{ST} by using cross-validation in terms of PCP. The estimation results will give τ and h_{ST} instead of λ, μ, and h_{ST} in this study.

The Nelder-Mead Simplex Method is employed to obtain τ and h_{ST}, which enable the model to attain the maximum PCP. This is a direct search method that does not use numerical or analytic gradients. If n is the length of x, a simplex in n-dimensional space is characterized by the $n + 1$ distinct vectors that are its vertices. In two-dimensional, a simplex is a triangle. During each step of the search, a new point in or near the current simplex is generated. The function value at the new point is compared with the function's values at the vertices of the simplex. Typically, one of the vertices is replaced by the new point, giving a new simplex. This step is repeated until the diameter of the simplex is less than the specified tolerance.

5.3.3 McNamara's Test

To demonstrate the effectiveness of the proposed model, the estimation results are compared with those obtained from the MNLM. In addition to the Percentage of Correctly Predicted (PCP), which measures the goodness-of-fit of the models, the difference between the accuracies achieved by the two methods (GTWLM and the conventional MNLM) is also assessed by McNamara's test (see also Foody, 2004). This test is based on the standardized normal test statistic:

$$
Z = \frac{f_{12} - f_{21}}{\sqrt{f_{12} + f_{21}}}
\tag{5.27}
$$

where f_{12} denotes the number of samples classified correctly and wrongly by the first and second models, respectively. Accordingly, f_{12} and f_{21} are the counts of classified samples, with which the first and second models disagree. A lower prediction error (higher accuracy) is identified by the value of Z. A negative value of Z indicates that the results from f_{12} are more accurate than the results from model f_{21}. At the commonly used 5% level of significance, the difference in the accuracies between the first and the second models is evaluated to be statistically significant if $|Z| > 1.96$.

5.4 Results and Discussion

The TWLM, GWLM, and GTWLM are performed by using the data of SEZ, and the estimates are reported in Tables 5.1, 5.2, and 5.3, respectively. Because the output of the local parameter estimates from WLM, GWLM, and GTWLM would be voluminous, only the minimum, median, and maximum value for each parameters are

Table 5.1 TWLM parameter estimate summaries

| Parameter | TWLR (bandwidth = 1.180; q = 700) | | | | | | | | |
| | Residential | | | Industrial | | | Transportation/commercial/others | | |
	Min	Med	Max	Min	Med	Max	Min	Med	Max
Constant	-1.97	0.905e-01	1.13	-3.59e-01	4.97e-01	2.46	2.18e-01	8.37e-01	2.66
Distance to commercial center	-1.40e-03	-4.95e-04	7.44e-04	-1.13e-03	-1.75e-04	7.26e-04	-2.03e-04	2.00e-04	1.05e-03
Distance to financial center	-4.49e-04	-1.26e-04	2.42e-04	-2.77e-04	-6.79e-05	7.26e-04	-4.79e-04	1.43e-05	3.43e-04
Distance to industrial center	-6.15e-04	-1.38e-04	1.22e-03	-1.79e-02	-1.17e-02	-4.62e-03	-5.60e-04	3.26e-04	5.36e-04
Distance to educational facilities	-1.14e-03	-6.59e-04	2.19e-04	-1.96e-04	1.65e-03	2.54e-03	-2.53e-04	9.21e-05	6.73e-04
Distance to railway infrastructure	-1.02e-04	-2.41e-05	3.25e-05	-3.21e-04	-1.87e-04	-6.09e-05	-2.14e-04	-1.26e-04	8.25e-05
Distance to road	-5.33e-03	-2.80e-03	2.66e-04	-1.68e-03	7.30e-04	7.39e-03	-9.54e-03	-3.42e-03	-3.97e-04
Population	-3.12e-06	3.69e-05	1.05e-04	-4.06e-06	6.97e-05	1.11e-04	1.89e-05	5.10e-05	1.08e-04
DEM	-2.47e-02	-1.19e-02	-9.51e-03	-3.76e-02	-2.19e-02	-5.23e-03	-2.52e-02	-1.19e-02	-6.57e-03
Slope	-1.37e-01	-6.87e-02	-4.56e-02	-2.66e-01	-8.16e-02	-6.85e-02	-1.04e-01	-6.52e-02	-3.79e-02
Planning for residential	1.49	2.45	3.83	7.06e-01	1.47	2.26	6.97e-01	8.00e-01	2.58
Planning for industrial	9.94	1.13e01	1.37e01	1.28	1.13e01	1.30e01	9.93	1.12e01	1.26e01
Planning for transportation/commercial/others	-1.40	-3.99e-02	9.35e-01	8.57e-01	1.30	2.18	3.81e-01	6.36e-01	2.08
No planning	-9.85e-01	-2.30e-01	1.60	-4.48e-01	1.47e-01	1.73	-3.07e-01	-7.34e-02	8.57e-01
PCP					75.4%				

Table 5.2 GWLM parameter estimate summaries

Parameter	GWLR (bandwidth = 0.5916; q = 1,323)								
	Residential			Industrial			Transportation/commercial/others		
	Min	Med	Max	Min	Med	Max	Min	Med	Max
Constant	-26.8e01	4.09	2.53e01	-3.29e01	10.1e01	6.22e01	-9.97	3.70	4.29e01
Distance to commercial center	-3.85e-02	-3.62e-04	6.20e-03	-2.35e-02	-3.68e-04	1.83e-02	-1.92e-02	1.06e-03	1.27e-02
Distance to financial center	-8.67e-03	-9.33e-04	1.38e-02	-8.79e-03	2.11e-04	2.41e-02	-6.77e-03	-5.21e-04	1.02e-02
Distance to industrial center	-1.55e-02	-2.59e-04	2.55e-02	-3.06	-7.09e-02	-4.16e-03	-8.77e-03	1.04e-03	1.75e-02
Distance to educational facilities	-5.35e-02	-1.58e-03	1.02e-02	-2.95e-02	4.28e-03	3.91e-02	-8.08e-03	2.40e-04	9.23e-03
Distance to railway infrastructure	-5.77e-03	-4.93e-04	6.78e-03	-2.81e-02	-1.17e-03	1.46e-03	-4.74e-03	-6.52e-04	7.22e-04
Distance to road	-8.11e-02	-9.96e-03	6.44e-03	-3.72e-02	3.26e-03	6.47e-02	-5.75e-02	-1.01e-02	9.74e-03
Population	-4.72e-04	2.52e-04	3.51e-03	-1.23e-03	1.69e-04	7.76e-03	-4.08e-04	4.16e-04	3.50e-03
DEM	-1.15	-1.08e-01	3.31e-03	-9.35e-01	-9.08e-02	5.24e-02	-3.48e-01	-6.92e-02	-1.50e-02
Slope	-3.16	-3.17e-01	3.74e-01	-5.49	-4.83e-01	2.03e-01	-2.87	-2.80e-01	7.39e-03
Planning for residential	-5.69e-01	9.64	1.05e02	-3.00e01	3.76	6.05e01	-1.71	3.34	6.21e01
Planning for industrial	-5.36	1.77e02	9.19e02	8.20	1.65e02	6.96e02	-1.28e01	1.73e02	9.85e02
Planning for transportation/commercial/others	-4.01e01	-5.18e-01	3.87e01	-9.59	4.60	3.69e01	-3.71	3.27	1.77e01
No planning	-5.05e01	-4.81e-01	3.12e01	-1.24e02	3.41	2.74e01	-6.17	1.40	1.77e01
PCP					79.2%				

Table 5.3 GTVLM parameter estimate summaries

Parameter	GTVLM (bandwidth = 0.6671; τ = 0.356; q = 920)								
	Residential			Industrial			Transportation/commercial/others		
	Min	Med	Max	Min	Med	Max	Min	Med	Max
Constant	−8.84e01	7.23	5.30e01	−5.48e01	1.38e01	6.99e01	−1.34e01	5.45	4.75e01
Distance to commercial center	−3.50e−02	−4.86e−04	1.24e−02	−2.88e−02	−2.95e−04	1.72e−02	−2.42e−02	2.93e−03	1.89e−02
Distance to financial center	−1.56e−02	−1.43e−03	2.22e−02	−1.76e−02	−2.06e−04	4.50e−02	−1.48e−02	−7.90e−04	2.04e−02
Distance to industrial center	−2.06e−02	−7.79e−04	3.30e−02	−5.10	−1.02e−01	−6.94e−03	−1.57e−04	7.31e−04	2.77e−02
Distance to educational facilities	−6.50e−02	−3.54e−03	2.62e−02	−3.49e−02	7.47e−03	6.90e−02	−1.23e−02	−1.79e−04	2.17e−02
Distance to railway infrastructure	−7.25e−03	−5.13e−04	8.75e−03	−2.99e−02	−2.20e−03	1.78e−03	−5.98e−03	−9.27e−04	1.58e−03
Distance to road	−9.41e−02	−1.63e−02	1.01e−02	−6.93e−02	8.17e−03	1.31e−01	−1.11e−01	−1.64e−02	6.78e−03
Population	−2.86e−03	3.15e−04	4.22e−03	−1.90e−03	5.91e−04	1.80e−02	−9.61e−04	6.01e−04	4.79e−03
DEM	−1.28	−1.47e−01	3.17e−02	−7.23e−01	−1.44e−01	2.10e−01	−5.09e−01	−8.26e−02	7.68e−02
Slope	−5.11	−5.02e−01	1.08	−5.77	−6.27e−01	9.54e−01	−3.73	−3.96e−01	1.28e−01
Planning for residential	−2.05e01	1.27e01	1.34e02	−4.04e01	3.81	7.43e01	−1.06e01	5.45	7.20e01
Planning for industrial	−7.74e01	1.56e02	6.91e02	−9.01	1.68e02	1.210e03	−5.99e01	1.67e02	7.46e02
Planning for transportation/commercial/others	−4.60e01	−1.11	7.99e01	−1.39e01	5.21	6.34e01	−1.50e01	5.06	2.38e01
No planning	−4.66e02	−1.06	9.60e01	−6.53e03	1.02	7.19e01	−1.10e01	1.64	3.26e01
PCP					82.3%				

Table 5.4 Comparison of PCPs between MNLM and GTWLM

Observed	MNLM						GTWLM					
	Predicted				Total	PCP	Predicted				Total	PCP
	0	1	2	3			0	1	2	3		
0	1,726	36	9	206	1,997	86.4%	1,798	13	5	161	1,977	90.9%
1	33	352	8	153	546	64.5%	25	370	7	144	546	67.8%
2	27	24	152	97	300	50.7%	2	3	248	47	300	82.7%
3	178	274	42	883	1,377	64.1%	136	178	28	1,035	1,377	75.2%
Total	1,964	686	211	1,339	3,113	74.1%	1,961	564	288	1,387	3,451	82.3%

Table 5.5 PCPs of TWLM and GWLM

Observed	TWLM						GWLM					
	Predicted				Total	PCP	Predicted				Total	PCP
	0	1	2	3			0	1	2	3		
0	1,738	27	14	198	1,977	87.9%	1,768	19	8	182	1,977	89.4%
1	34	302	10	200	546	55.3%	27	355	15	149	546	65.0%
2	16	16	168	100	300	56.0%	4	3	236	57	300	78.6%
3	174	197	46	960	1,377	69.7%	151	220	38	968	1,377	70.3%
Total	1,962	542	238	1,458	3,168	75.4%	1,950	597	297	1,356	3,327	79.2%

provided to give a summary of the distribution. In TWLM, GWLM, or GTWLM, the magnitude of all the parameters in the global models are between the minimum and the maximum, and the signs of the median for all the parameters are almost the same as the MNLM. The balance factor between the spatial distance effect and the temporal distance effect τ being 0.356 implies the leading role played by the spatial effect.

Table 5.4 provides the comparison of PCPs between the MNLM and the GTWLM. It should be noted that the PCP has increased from 74.1% in the MNLM to 82.3% in GTWLM. As seen from Table 5.5, the PCP is 75.4% in TWLM and 79.2% in GWLM. A comparison with the PCP indicates that GTWLM gives a better fit of data than the TWLM, GWLM, and MNLM because GTWLM can handle both spatial and temporal non-stationarity. Moreover, Table 5.5 shows that the GWLM achieved a better goodness-of-fit than that of TWLM model. It reveals that the temporal non-stationary effect is less significant than that of spatial non-stationarity. A possible reason is that the experimental data include only the seven time spots, and these provide less information.

TWLM, GWLM, and GTWLM show improvements over the MNLM in terms of PCP. However, it is still necessary to investigate whether those models perform significantly better than MNLM from a statistical viewpoint. The improvement on TWLM and GWLM by GTWLM should be also tested.

For comparison purposes, McNamara's test is employed to test the significant difference between MNLM, TWLM, GWLM, and GTWLM, and the results are

Table 5.6 Significance comparison for GWR-based models

Models Comparison	Z values		
	TWLM	GWLM	GTWLM
MNLM	−2.64	−9.57	−14.16
TWLM	−	−6.73	−12.08
GWLM	−	−	−7.75

Table 5.7 Comparison of MNLM, TWLM, GWLM, and GTWLM with the Kappa coefficients

	MNLM	TWLM	GWLM	GTWLM
Kno	0.65	0.67	0.72	0.76
Klocation	0.63	0.64	0.69	0.73
Kquantity	0.90	0.94	0.97	0.98

shown in Table 5.6. They clearly indicate (negative value) that the GTWLM model performed better than GWLM and MNLM. The Z values between TWLM and GWLM are −5.33 and −13.67, respectively, thereby indicating that GWLM substantially outperformed TWLM. Also, the Z value between TWLM and MNLM is −2.64, which is less than −1.96. These results demonstrate a significant difference between the MNLM, TWLM, GWLM, and TWGR models at the 95% confidence level. These comparisons evince that GTWLM outperforms MNLM, GWLM, and TWLM in the model accuracy.

Their Kappa measures are also calculated and shown in Table 5.7. All the Kappa coefficients show that GTWLM provides a better result than the other models.

Furthermore, the local parameter estimates of TWLM, GWLM, and GTWLM, which denote local relationships, can be displayed visually. Considering the coefficients of "Planning for residential in land use residential" as an example, they can be grouped into several groups and each group can be colored to visualize the spatial variation patterns of this parameter.

The spatial distributions of the parameter estimates for "Planning for residential in land use residential" of TWLM, GWLM, and GTWLM are shown in Figure 5.2. For the TWLM, no significant spatial variation is observed over time. The spatial variations of "Planning for residential in land use residential" in both GWLM and GTWLM share analogous distributions. This, however, excludes the spatial variation in GTWLM, which portrays heterogeneity in greater detail. It can be also inferred that the spatio-temporal non-stationarity of the GTWLM is dominated by the spatial effect for this land use dataset.

Figure 5.2 illustrates that the parameter "Planning for residential in land use residential" in GTWLM is smaller in Nanshan and Yantian. This holds true especially in the north-western part of Nanshan, which is next to Baoan, and bigger in Futian and Luohu. This serves to indicate that the planning for residential areas is implemented better in the center of the SEZ. Owing to the rapid development, planning for

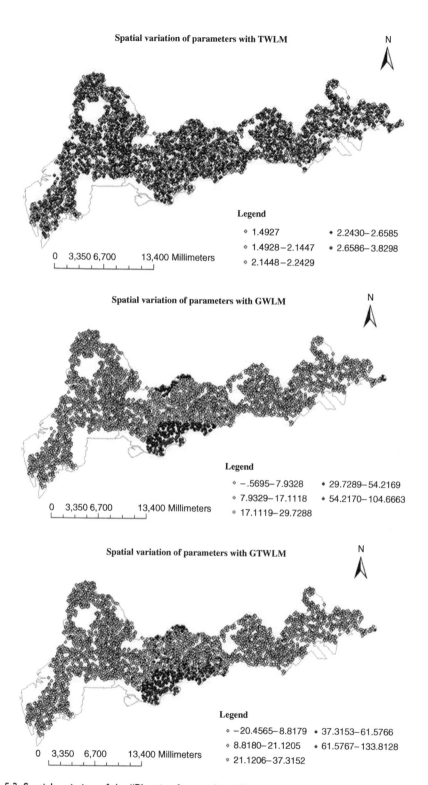

Figure 5.2 Spatial variation of the "Planning for residential" coefficient in residential

residential almost does not work in some places for, *e.g.*, the region in the north-west part of Nanshan.

5.5 Conclusion

The local models that consider non-stationarity refute the criticism that those adopting a quantitative approach to investigate the land use change process are concerned only with the search for broad generalizations and have little interest in identifying local exceptions. Local forms of analysis also provide the relation between the outputs of the spatial techniques and the powerful visual display capabilities of GIS. Notably, they provide detailed information on spatial and temporal relationships, and this enhances the model development process and facilitates understanding of the land use change.

In this chapter, GWLM considering spatial non-stationarity has been extended to GTWLM in a spatio-temporal framework. The land use change in SEZ, Shenzhen has been modeled. Compared with the other models, the GTWLM model shows a significant improvement in the percentage of correctly predicted items. Compared with the global MNLM, the PCP values of TWLM and GWLM increase from 74.1% to 75.4% and 79.2%, respectively. The GTWLM yielded a considerably higher PCP of 82.3%. Statistical tests evince that a significant difference exists between MNLM, TWLM, GWLM, and GTWLM. The Kappa coefficients also indicate that the GTWLM is better than the other models. Consequently, it can be safely concluded that it is meaningful to incorporate spatial and temporal non-stationarity into a land use change model.

However, several limitations remain to be studied in future. For instance, only seven temporal periods are available. The inadequacy in temporal information can be expected to degrade the model performance of TWLM and GTWLM. There exists a possibility that unobservable factors (not covered by our dataset) and the individual effect of land cells could influence the estimation results. More efficient spatio-temporal distance and weighting functions should be designed to constitute the weight. Besides, the spatio-temporal autocorrelation also should be considered in a more flexible spatio-temporal framework. Hence, a more elaborate version of GTWLM is covered in Chapter 6.

Chapter 6

Spatio-temporal Panel Logit Model

Anselin (1988, 1998) provides three principal methods for addressing the spatial autocorrelation: use of spatial stochastic processes; a direct representation of autocorrelations; and a non-parametric framework. The second method is preferred in our context, and it is extended to consider the spatio-temporal autocorrelation in land use change models. Meanwhile, the individual effect can be considered in a panel data framework. The combination of these two methods allows the consideration of both the spatio-temporal autocorrelation and the individual effect in land use change modeling.

In this chapter, two improvements (spatio-temporal autocorrelation and individual effect) in the MNLM are discussed initially. Then, the panel concept is introduced. Based on the panel data analysis framework, ST-PLM is developed. The implementation and evaluation of the proposed model are discussed in detail. Subsequently, results showing the enhanced performance of the model are discussed. Finally, the conclusion summarizing the proposed model is provided.

6.1 Introduction

Spatial autocorrelation refers to the tendency of spatial variables and spatial land use data to be dependent (Legendre & Fortin, 1989). Spatial autocorrelation, exhibited by substantial amounts of spatial variables and land use data, has been studied by numerous researchers. The Spatial Autologistic Regression model or the Spatial AutoLogit (SAL) model, the logistic regression model incorporating spatial autocorrelation, has been devised (Dubin, 1995; Dubin, 1997; LeSage, 1998). Such models have proven to be effective in regression analysis with the consideration of spatial autocorrelation (Páez & Suzuki, 2001). Similar to the spatial lag model, a spatial error model (Anselin, 1988) can be employed to deal with spatial autocorrelation in a regression model.

On the other hand, less attention has been paid to temporal autocorrelation. Land use change modeling entails considering both the spatial and the temporal autocorrelation. Otherwise, inefficient parameter estimates and inaccurate measures of statistical significance will result. Two recent studies in land use change analysis have taken into consideration the spatial and temporal autocorrelation in their statistical models. An and Brown (2008) introduce the concepts and models in survival analysis, and their potential applications in land use change. While survival analysis has proven to be effective in addressing temporal complexities, their model does not consider spatial autocorrelation usually based on the spatial weight structure. The land use model generated by Huang *et al.* (2009a) accounts for spatial neighbors in the last cross-section and employs a modified exponential smoothing technique to produce a smoothed model from a series of bi-temporal spatial models for different time periods. Although spatial and temporal autocorrelation is partly considered in this model, it does not identify the individual specific effect in each cell.

In view of the shortcomings of the aforementioned models, this chapter aims to construct an innovative model to delineate the spatio-temporal pattern changes of land use. A panel data model, covered in detail in Section 5.2, can be used to represent the individual effect of a land cell. Thus, a panel data framework is employed to model land use change in this study. Based on the panel data analysis framework, the proposed ST-PLM considers the random effects of a land use cell with reference to both space and time and establishes the relationship between various factors and land use patterns over time. Considering spatial and temporal autocorrelation, the model incorporates the covariance of a cell to other cells into the model formulation. To incorporate the individual effect, the model accounts for the land use type of a cell in the initial year 1990. Notably, the proposed model integrates spatial autocorrelation with temporal autocorrelation in the same framework and performs the maximum simulated likelihood estimations. A study of spatio-temporal land use change in the SEZ has been carried out to demonstrate the superior performance of the proposed model over the MNLM.

6.2 Panel Data Model

A panel that is a cross-section or group of objects periodically observed over a given time span (Anselin, 1988; Greene, 2000) can be used to represent the multi-temporal land use distributions of a region. Panel data analysis thus serves as a suitable tool for spatio-temporal land use change modeling.

6.2.1 Panel Data

A panel dataset follows a given sample of individuals over time, and subsequently provides multiple observations on each individual in the sample. A sample panel dataset, consisting of a pool of observations on a cross-section over four time periods, is presented in Figure 6.1.

Hsiao (2003) describes several major advantages of panel datasets over conventional cross-sectional or time-series datasets. Panel data usually offer the researchers a large number of data points. This serves to increase the degrees of freedom and to reduce the collinearity between the explanatory variables, hence improving the efficiency of the econometric estimates. More importantly, panel data allow a researcher to analyze individual effect.

6.2.2 Panel Data Model

A panel data model, typically handling all the time series and cross-sectional data, can be employed for spatio-temporal analysis (Greene, 2000; Wooldrige, 2002). Grid cells

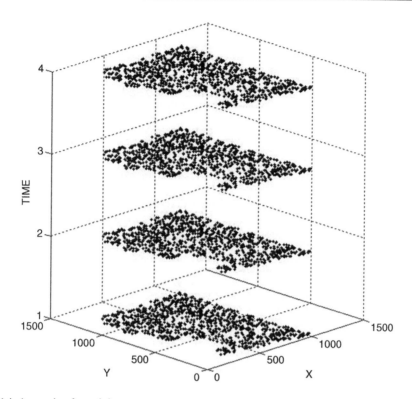

Figure 6.1 A sample of panel dataset

in panel data involve a minimum of two dimensions (a cross-sectional dimension, indicated by subscript *i*, and a time series dimension, indicated by subscript *t*). The formula of the panel data model is:

$$y_{it} = X'_{it}\beta + u_{it}, i = 1, 2, \ldots, N, t = 1, 2, \ldots, T \tag{6.1}$$

where X_{it} and β are vectors of explanatory variables and parameters, respectively, and u_{it} is an error component whose assumption is different from the classic assumption in linear regression. The panel data model in one-way error component assumes the following form:

$$u_{it} = \mu_i + v_{it} \tag{6.2}$$

where $\mu_i \sim IID(0, \sigma_\mu^2)$ and $v_{it} \sim IID(0, \sigma_v^2)$. μ_i is used to identify unobserved individual effect. If μ_i is assumed to be a constant parameter to be estimated, the model will become a panel data model with fixed effect. To account for spatio-temporal auto-correlation by using a direct representation in error component in land use change analysis, the error part should be allowed to vary randomly across land use choices. This idea led to a panel data model with random effects where choice probabilities for repeated observations on the same individual share the same unobserved random effects (Train, 2003).

6.2.3 Estimation of Panel Data Model with Random Effect: Maximum Simulated Likelihood Estimation (MSLE)

Because there is no closed form solution to the marginal likelihood for the panel data model with random effect, simulation-based methods are used to integrate the random latent heterogeneity term: error part. This procedure is similar to the maximum likelihood (ML) except that the simulated probabilities are used instead of the exact probabilities.

In ML, the log-likelihood function is as follows:

$$LL(\beta) = \sum_i \ln P_i(\beta) \tag{6.3}$$

where β is a vector of parameters, $P_i(\beta)$ is the (exact) probability of the observed choice of observation i, and the summation is over a sample. β can be estimated by maximizing the log-likelihood function. If let $\hat{P}_i(\beta)$ be a simulated approximation to $P_i(\beta)$, the simulated log-likelihood function is:

$$SLL(\beta) = \sum_i \ln \hat{P}_i(\beta) \tag{6.4}$$

β can also be estimated by maximizing the simulated log-likelihood function. The properties of MSL were derived by Gourieroux and Monfort (1993). Train (2003) and Greene (2000) provide the technical details of the likelihood function and methods to maximize the likelihood function based on simulation techniques.

6.3 Methodology

6.3.1 Formulation of STLM

The general form of the MNLM used for land use change analysis is as follows:

$$prob(y_i^j = 1) = \frac{\exp(\beta^j x_i)}{\sum_{j=0}^{J} \exp(\beta^j x_i)}, \, j = 0, \ldots, J \tag{6.5}$$

where y_i^j is an indicator of the observed land use type j for cell i: if yes, y_i^j equals 1; if no, y_i^j equals 0. J is the set of all the land use types. In this case, $j = 0$ denotes undeveloped, 1 denotes residential, 2 denotes industrial, and 3 denotes transportation /commercial/ others. β^j and x_i are the vectors of parameters for type j and explanatory variables for cell i, respectively. x_i refers to the variables of the driver factors listed in Table 2.2 for cell i in this analysis.

$$\beta^j x_i = \beta_1^j \text{distance to Commercial Center}_i + \beta_2^j \text{distance to Financial Center}_i + \ldots +$$
$$\beta_{13}^j \text{No Planning}_i \tag{6.6}$$

In order to reduce the number of parameters of the model, a value of 0 is assigned to all the variables of undeveloped land (base category), i.e., $\beta^0 = 0$.

The proposed model also aims to incorporate individual effect. Following the framework of the panel data model, Equation 6.5 is revised as follows to identify the individual effect of a sampled cell:

$$prob(y_i^j = 1) = \frac{\exp(\beta^j x_i + \mu_i^j)}{\sum_{j=0}^{J} \exp(\beta^j x_i + \mu_i^j)} \quad, j = 0, \ldots, J \tag{6.7}$$

where μ_i^j represents an individual random effect. y_i^j is an indicator of the observed land use type for cell i.

Considering the context of land use change, a difference in land use exists between the built-up area and the undeveloped area. μ_i^j is assumed to be normally distributed with mean $\alpha_0^j + \alpha_1^j \left(y_{i0}^{j=1} + y_{i0}^{j=2} + y_{i0}^{j=3} \right)$ and a standard deviation δ^j as in the following formula:

$$\mu_i^j = \alpha_0^j + \alpha_1^j \left(y_{i0}^{j=1} + y_{i0}^{j=2} + y_{i0}^{j=3} \right) + \delta^j \varepsilon_i^j \tag{6.8}$$

where ε_i^j is assumed to be independent and identically distributed (*iid*) *Standard Normal Distribution* across the land use types and observations. α_0^j, α_1^j, and δ^j are the parameters to be estimated. y_{i0}^j is the indicator of the observed land use type j for cell i in the first year 1990. In Equation 6.8, if ε_i^j is assumed to be *iid* across land use types and observations, this model will then be similar to an MNLM for panel data with random effects (Greene, 2000).

Spatio-temporal autocorrelation exists in the change process and it should be incorporated into the proposed model. Thus, an individual random effect is used to identify the spatio-temporal autocorrelation between the cells, which means the covariance matrix of ε_i^j is no longer assumed to be independent and identically distributed (*iid*) *Standard Normal Distribution*. The autocorrelation between cells i and m is assumed to be inversely proportional to the spatio-temporal distance between them. This is expressed as follows:

$$corr\left(\varepsilon_i^j, \varepsilon_m^j\right) \propto \exp\left(\frac{-d_{im}^{ST}}{\eta}\right) \tag{6.9}$$

Where d_{im}^{ST} is the spatio-temporal distance between cells i and m, and η is a scale factor. If there is a total of n observations, then the covariance matrix can be defined as follows:

$$corr(\varepsilon^j, \varepsilon^j) = \begin{pmatrix} \exp\left(\frac{-d_{11}^{ST}}{\eta}\right) & \exp\left(\frac{-d_{12}^{ST}}{\eta}\right) & \cdots & \exp\left(\frac{-d_{NN}^{ST}}{\eta}\right) \\ \exp\left(\frac{-d_{21}^{ST}}{\eta}\right) & \exp\left(\frac{-d_{22}^{ST}}{\eta}\right) & \cdots & \exp\left(\frac{-d_{2n}^{ST}}{\eta}\right) \\ \vdots & \vdots & \ddots & \vdots \\ \exp\left(\frac{-d_{n1}^{ST}}{\eta}\right) & \exp\left(\frac{-d_{n2}^{ST}}{\eta}\right) & \cdots & \exp\left(\frac{-d_{nn}^{ST}}{\eta}\right) \end{pmatrix} \tag{6.10}$$

The spatio-temporal weighted function used in GTWLM is adopted to identify $\exp(\frac{-d_{im}^{ST}}{\eta})$. Thus, $\exp(\frac{-d_{im}^{ST}}{\eta})$ can be divided into two parts: $\exp(\frac{-d_{im}^{S}}{b_S^2})$ and $\exp(\frac{-d_{im}^{T}}{b_T^2})$.

$$\exp\left(\frac{-d_{im}^{ST}}{\eta}\right) = \exp\left(\frac{-d_{im}^{S}}{b_S^2}\right) * \exp\left(\frac{-d_{im}^{T}}{b_T^2}\right) \tag{6.11}$$

where d_{im}^{S} and d_{im}^{T} are the spatial and the temporal distance, respectively. The same b_S^2 and b_T^2 as GTWLR are used here.

Overall, the ST-PLM's log-likelihood function is formulated as follows:

$$\ln(L) = \sum_{i=1}^{n} \sum_{j=0}^{J} y_i^j \ln\left(p\left(y_i^j = 1 | x_i; u_i^j, \beta^j\right) f\left(u_i^j | y_{i0}, \varepsilon_i^j\right)\right) \tag{6.12}$$

According to Equation 6.9, u_i^j can be obtained by parameter ω^j, which consists of α_0^j, α_1^j, and δ^j. Equation 6.12 can be rewritten as follows:

$$\ln(L(\beta^j, \omega^j)) = \sum_{i=1}^{n} \sum_{j=0}^{J} y_i^j \ln\left[\int_{\mu_i^j} \left\{\frac{\exp(\beta^j x_i + \mu_i^j)}{\sum_{j=0}^{J} \exp(\beta^j x_i + \mu_i^j)}\right\} g(\mu_i^j | \omega^j) \partial \mu_i^j\right] \tag{6.13}$$

Theoretically, the parameters of the model are estimated by maximizing the log-likelihood function. The computation of the probabilities for each land use type has posed significant problems in this context. The log-likelihood function above involves the estimation of an integral, which cannot be evaluated analytically for its lack of a closed-form solution. Considering the speed and precision, simulation techniques (Train, 2003) have been used to estimate the log likelihood function. The integrals in the choice probabilities are approximated by the Monte Carol technique. Then, the resulting simulated log-likelihood function is maximized. Therefore, the MSL is adopted (same as ML, except that the simulated probabilities are used instead of the exact probabilities).

Initially, three (for j = 1, 2, and 3) n-dimensional normally distributed random vectors should be generated with the autocorrelation matrix given by Equation 6.9. In order to obtain the vectors, Equation 6.9 should be decomposed to a lower-triangular matrix to be multiplied by n-dimensional normally distributed random vectors. Frequently, the Cholesky decomposition method is adopted for simulating systems with multiple correlated variables (the intervariable autocorrelations matrix is decomposed to obtain the lower-triangular L. Applying this to a vector of uncorrelated simulated shocks, v produces a shock vector $L*v$ with the covariance properties of the system modeled). However, the autocorrelation matrix between the observations can be guaranteed only to be symmetric, not positive definite. This makes the application of the Cholesky decomposition method impossible. Instead, the generalized Cholesky decomposition method devised by Gill and King (2004) is used. In the Cholesky decomposition method, when a symmetric matrix is not positive definite, in order to produce a positive definite matrix, a nonnegative diagonal matrix with element values as small as possible is added to the original matrix. For more details about the generalized Cholesky decomposition method, please refer to Appendix A.

The simulated probabilities for this case are obtained from the following steps.

Step 1) Decompose the covariance matrix into a triangular matrix by the generalized Cholesky decomposition method.

Step 2) Obtain *three* (for j = 1, 2, and 3) sequences of draws following the *iid* normal distribution by shuffled Halton draws (please refer to Appendix B for details). Each sequence consists of *n* number for each cell.

Step 3) Multiply the lower-triangular matrix by *n iid* normal values and obtain two new sequences.

Step 4) Calculate the Equation 6.8 in accordance with ε_i^j obtained from the new sequences.

Step 5) Repeat steps 2 to 4 for 100 times and average the results.

The integrals in the choice probabilities are approximated by the steps. Then, the resulting simulated log-likelihood function is maximized. The model's parameters are estimated finally.

6.3.2 Evaluation of the ST-PLM's Improvement: Akaike's Information Criterion (AIC)

Akaike's information criterion is grounded in the concept of entropy, effectively offering a relative measure of the information lost when a given model is used to describe reality. This describes the tradeoff between bias and variance in model construction or that between precision and complexity of the model.

In the general case, the AIC can be calculated as follows:

$$AIC = 2k - 2\ln(L) \tag{6.14}$$

where k is the number of parameters in the model, and L is the maximized value of the likelihood function for the estimated model. Given a dataset, several competing models may be ranked in accordance with their AIC (the lowest AIC represents the best). The AIC methodology attempts to find the model that best explains the data with a minimum of free parameters. In this case, AIC is employed to decide which model is better (*i.e.*, MNLM *vs* ST-PLM).

6.4 Results and Discussions

The following section explains the model validation in this case. The last five years map data (1996, 2000, 2002, 2004, 2006, and 2008) are used for the estimation of the model. The data of the year 1990 are employed to construct the mean of μ_{it}^j. For each sample cell, the probabilities on all the land use types are computed with the ST-PLM. Then, the probabilities are compared and the land use type with the highest probability is selected for the cell. Table 6.1 presents the parameters generated by the ST-PLM by using the MSL estimation developed in this study. The log likelihood is –2,323.4 and the PCP is 79.4%.

Table 6.1 Estimation results of ST-PLM

Parameters	Coef. (Residential)	t-stats	Coef. (Industrial)	t-stats	Coef. (Commercial/ Transportation/ others)	t-stats
Distance to commercial center	−2.08e-4	−9.62e-1	−2.37e-4	−1.12	3.13e-4	2.43
Distance to financial center	−2.20e-4	−1.60	−7.41e-05	−6.09e-1	−4.74e-05	−5.95e-1
Distance to industrial center	5.20e-05	2.55e-1	−9.65e-3	−1.36e01	1.35e-4	0.824
Distance to educational facilities	−7.89e-4	−2.73	1.39e-3	5.13	−1.21e-4	−0.697
Distance to railway infrastructure	−7.07e-05	−1.81	−1.79e-4	−3.84	−9.06e-05	−2.90
Distance to road	−2.44e-3	−3.90	2.17e-4	5.81e-1	−1.20e-3	−4.64
Population	4.04e-05	3.03	4.08e-05	2.50	5.10	4.40
DEM	−1.01e-2	−2.81	−1.89e-2	−3.09	−9.07e-3	−5.33
Slope	−6.67e-2	−5.01	−8.91e-2	−5.16	−5.62e-2	−7.37
Planning for residential	2.24	7.87	1.10	3.12	8.91e-1	3.75
Planning for industrial	1.29e01	7.22e-2	1.28e01	7.17e-2	1.29e01	7.22e-2
Planning for Transportation/ commercial/others	−1.21e-1	−4.43e-1	1.10	3.53	8.23e-1	4.47
No planning	−9.23e-3	−3.06e-2	1.25e-1	3.67e-1	3.41e-1	1.83
α_0^j	4.09e-1	1.20	1.10	2.70	−2.51e-1	−1.00
α_1^j	5.62e-1	2.65	2.21e-1	8.95e-1	2.89	1.64e01
δ^j	1.44	5.39e-1	−6.12	−1.81	4.73	2.06
Log likelihood function			−2.3234e+03			
PCP			79.4%			

It can be found that δ^j, which represents spatio-temporal autocorrelation, is significant for industrial and commercial/transportation/others with a large absolute t-statistics. The positive parameters of individual effect for all the built-up area reveals built-up land use types have persistently featured.

The comparison between MNLM and ST-PLM is used to examine whether the proposed ST-PLM offers any improvement. Table 6.2 compares the accuracies achieved by MNLM and ST-PLM. ST-PLM incorporates the spatio-temporal auto-correlation and individual effect in one land use dataset, whereas MNLM ignores it. Consequently, the spatio-temporal model achieves a higher overall PCP (79.4%) than MNLM (74.1%). Also, the accuracies for undeveloped, industrial, and commercial/transportation/others achieved by ST-PLM are better (*i.e.*, 86.4% *vs* 90.4%; 50.7% *vs* 56.0%; 64.1% *vs* 76.6%). Because the distribution of residential is dispersed, the

Table 6.2 Comparison of PCPs between MNLM and ST-PLM

Observed	MNLM (Multinomial logit model)						ST-PLM (Spatio-temporal panel logit model)					
	Predicted				Total	PCP	Predicted				Total	PCP
	0	1	2	3			0	1	2	3		
0	1,726	36	9	206	1,997	86.4%	1,806	35	15	121	1,997	90.4%
1	33	352	8	153	546	64.5%	54	305	26	161	546	55.9%
2	27	24	152	97	300	50.7%	28	22	168	82	300	56.0%
3	178	274	42	883	1,377	64.1%	187	86	49	1,055	1,377	76.6%
Total	1,964	686	211	1,339	3,113	74.1%	2,075	448	258	1,419	3,334	79.4%

Table 6.3 Comparison of MNLM and ST-PLM with the Kappa coefficients

	MNLM	ST-PLM
Kno	0.65	0.73
Klocation	0.63	0.71
Kquantity	0.90	0.90

consideration of spatio-temporal autocorrelation is not viable for this land use type. Thus, the accuracy for residential in MNLM is lower than that for ST-PLM.

To further examine whether ST-PLM demonstrates significant improvements over MNLM, McNamara's test and AIC test are performed. For McNamara's test, the Z value calculated is –8.3590 (its absolute value is greater than 1.96). Hence, it can be concluded that at a 5% level of significance, the difference in the accuracies between the two models is statistically significant. In other words, ST-PLM notably outperforms MNLM. For the AIC test, the values of MNLM and ST-PLM are 5,424 and 4,678.8, respectively. This reveals that the ST-PLM better explains the data with fewer free parameters.

The Kno in Table 6.3 evinces that the ST-PLM consistently achieves a better result than MNLM. Specifically, ST-PLM's ability to identify location is better than MNLM's, and the two models have the same ability when it comes to specifying quantity.

6.5 Conclusion

Over the years, spatial autocorrelation has been seriously considered in the context of land use change, which is inherently a spatio-temporal process. A natural step forward involves incorporating both spatial and temporal autocorrelation to analyze the change.

This chapter has presented such an endeavor using ST-PLM, which provides a powerful option to establish the relationship between the explanatory variables identified and the land use type. Notably, the proposed model effectively explores the spatial and temporal autocorrelation. In the proposed model, the spatio-temporal autocorrelation is considered in the random effect component ε_{it}^j with an assumption that the autocorrelation between ε_{it}^j is inversely proportional to the spatio-temporal distance between them. The proposed model has been validated by using the SEZ, Shenzhen land use data. The case study demonstrates that the proposed model can improve the PCP as well as the accuracy. A McNamara's test and an AIC test were performed, which corroborate the superior performance of ST-PLM over the MNLM. Besides, the Kno also shows that ST-PLM consistently provides a better result than MNLM. Specifically, ST-PLM's ability to specify location is better than MNLM's, and the two models have the same ability in specifying quantity.

As demonstrated by the model developed in this study, individual effect and spatio-temporal autocorrelation should be integrated when studying the dynamics of land use. A rigorous structure for spatio-temporal autocorrelation in conjunction with a solid statistical model can serve as an effective alternative for understanding land use change patterns.

Chapter 7

Generalized Spatio-temporal Logit Model

Non-stationarity and autocorrelation are two closely-related issues. Many evidences point to the existence of non-stationarity in the presence of autocorrelation and vice-versa. It is hence customary to consider both of them in a single model. Besides, individual effect is also a very important issue in land use change modeling. An ideal model for land use change study should solve the existing problems as much as possible. As discussed earlier, the GTWLM and ST-PLM can handle those issues, respectively. Thus, the integration of GTWLM and ST-PLM holds immense potential in solving the major challenges in this domain.

This chapter firstly introduces two important problems in spatio-temporal data modeling: spatio-temporal non-stationarity, and spatio-temporal autocorrelation, and it briefly summarizes the earlier work. Then, to construct and estimate both the problems simultaneously, the GTWLM is integrated with the ST-PLM to be GSTLM. Then, the chapter offers the detailed estimated technique for the GSTLM with MSLE and the generalized Cholesky decomposition procedure. In Section 7.4, the study of land use change by GSTLM in SEZ, Shenzhen is performed and reported. Subsequently, a spatio-temporal analysis of land use in SEZ, Shenzhen is conducted and the conclusions are drawn accordingly.

7.1 Introduction

Statistically modeling the quantitative relationships between the response variable and the explanatory variables for spatio-temporal data involves two important problems. The first problem is that spatial or temporal non-stationarity occurs in the relationships being modeled. The second problem involves the spatial or temporal autocorrelation that exists between the observations. Traditional models such as MNLM used in land use change modeling have largely ignored these two issues that violate the basic assumptions. Spatio-temporal non-stationarity violates the assumption that a single relationship exists across the sample data observations, whereas spatio-temporal auto-correlation violates the assumption that explanatory variables are fixed in repeated sampling.

Therefore, alternative approaches have been developed to model the aforementioned. For non-stationarity, one of the well-known models is the nonparametric local linear regression introduced by Mcmillen (1996) and Mcmillen and McDonald (1997). Brunsdon *et al.* (1996) labeled them geographically weighted regression

(GWR). GWR allows the exploration of the variation of the parameters, and it therefore has received considerable attention (Brunsdon *et al.*, 1996; Fotheringham & Brunsdon, 1999; Fotheringham *et al.*, 2001; Huang & Leung, 2002; Yu & Wu, 2004). For autocorrelation, Ord (1975) proposed the use of autoregressive and moving average terms in regression models to account for spatial autocorrelation in the response variable and the residuals, respectively. Cressie (1993) addressed spatial effects by modeling the residual variance–covariance matrix directly. Chapters 5 and 6 discuss the extension of the models into a spatio-temporal framework and developed the GTWLM and the ST-PLM to deal with spatio-temporal non-stationarity and autocorrelation in land use change modeling, respectively.

However, most of previous work deals with the aforementioned problems on an individual basis. Even though the two problems are theoretically distinct, many evidences show that there exists non-stationarity in the presence of autocorrelation, and an inadequate model that fails to capture non-stationarity will result in residuals that exhibit autocorrelation. Lesage (2005) argued for covering both the spatial and the temporal heterogeneity and autocorrelation effects in a mixed model. In fact, non-stationarity and autocorrelation share some similarities. For instance, both depend on the definition of spatio-temporal weighting matrices and both deal with spatio-temporal dependencies in data. Besides, the spatio-temporal weighting function in GTWLM and the spatio-temporal error autocorrelation matrix adopt the same form to quantify the notion of proximity. Although spatio-temporal non-stationarity and autocorrelation are often related in the context of modeling, we were able to find only a few studies that jointly construct and estimate both of them simultaneously. Therefore, building an integrated model with spatio-temporal non-stationarity and autocorrelation effects assumes research significance. Brunsdon *et al.* (1998) attempted to model both spatial non-stationarity and spatial autocorrelation in a complicated process. They proposed a geographically weighted regression with spatially lagged objective variable mode (GWRSL) to exploit the spatial variations. But their model ignores temporal information and individual effect.

This chapter ascertains the links between the two problems and focuses on integrating GTWLM and ST-PLM. A GSTLM is developed to utilize the spatio-temporal effect in land use change in SEZ, Shenzhen.

7.2 Generalized Spatio-temporal Logit Model (GSTLM)

The GSTLM is expressed as follows:

$$\log\left(\frac{prob(y_i^j = 1)}{prob(y_i^0 = 1)}\right) = \sum_p \beta_p^j(u_i, v_i, t_i)X_{ip} + \mu_i^j, \, j = 0, \ldots, J \tag{7.1}$$

where y_i^j is an indicator of the observed land use type j for cell i: if yes, y_i^j equals 1; if no, y_i^j equals 0. J is the set of all the land use types. Herein, $j = 0$ denotes undeveloped, 1 denotes residential, 2 denotes industrial, and 3 denotes transportation /commercial/ others. $\beta_p^j(u_i, v_i, t_i)$ are p continuous functions of the spatio-temporal coordinates (u_i, v_i, t_i) in the study area. X_{ip} refers to the p variable of driver factors listed in Table 2.2 for cell i in this analysis. Following the form of ST-PLM, μ_i^j, which represents the individual random effect, is assumed to be normally distributed. The mean $\alpha_0^j + \alpha_1^j\left(y_{i0}^{j=1} + y_{i0}^{j=2} + y_{i0}^{j=3}\right)$ and standard deviation δ^j are represented as follows:

$$\mu_i^j = \alpha_0^j + \alpha_1^j\left(y_{i0}^{j=1} + y_{i0}^{j=2} + y_{i0}^{j=3}\right) + \delta^j \varepsilon_i^j \tag{7.2}$$

where y_{i0}^j is the indicator of the observed land use type j for cell i in the first year 1990. To identify the spatio-temporal autocorrelation between the cells, the covariance matrix of ε_i^j is no longer an identity matrix. The autocorrelation between cell i and cell m is assumed to be inversely proportional to the spatio-temporal distance between them. This can be expressed as follows:

$$corr\left(\varepsilon_i^j, \varepsilon_m^j\right) \propto \exp\left(\frac{-d_{im}^{ST}}{h_{ST}}\right) \tag{7.3}$$

where h_{ST} is the bandwidth to be estimated, and d_{im}^{ST} is the spatio-temporal distance between cells i and m,

$$d_{im}^{ST} = \sqrt{\lambda((u_i - u_m)^2 + (v_i - v_m)^2) + \mu(t_i - t_m)^2} \tag{7.4}$$

λ and μ are scale factors to balance the different effects used to measure the spatial and temporal distance in their respective metric systems. If there are n observations in total, then the covariance matrix can be expressed as follows:

$$corr(\varepsilon^j, \varepsilon^j) = \begin{pmatrix} \exp\left(\frac{-d_{11}^{ST}}{h_{ST}}\right) & \exp\left(\frac{-d_{12}^{ST}}{h_{ST}}\right) & \cdots & \exp\left(\frac{-d_{1n}^{ST}}{h_{ST}}\right) \\ \exp\left(\frac{-d_{21}^{ST}}{h_{ST}}\right) & \exp\left(\frac{-d_{22}^{ST}}{h_{ST}}\right) & \cdots & \exp\left(\frac{-d_{2n}^{ST}}{h_{ST}}\right) \\ \vdots & \vdots & \ddots & \vdots \\ \exp\left(\frac{-d_{n1}^{ST}}{h_{ST}}\right) & \exp\left(\frac{-d_{n2}^{ST}}{h_{ST}}\right) & \cdots & \exp\left(\frac{-d_{nn}^{ST}}{h_{ST}}\right) \end{pmatrix} \tag{7.5}$$

In line with GTWLM, α_0^j, α_1^j, and δ^j, which are the fixed parameters to be estimated, are also allowed to vary over space and time, as shown below:

$$\mu_{it}^j = \alpha_0^j(u_i, v_i, t_i) + \alpha_1^j(u_i, v_i, t_i)\left(y_{i0}^{j=1} + y_{i0}^{j=2} + y_{i0}^{j=3}\right) + \delta^j(u_i, v_i, t_i)\varepsilon_i^j \tag{7.6}$$

Thus, Equation 7.1 can be revised as follows:

$$\log\left(\frac{prob(y_i^j = 1)}{prob(y_i^0 = 1)}\right)$$

$$= \sum_p \beta_p^j(u_i, v_i, t_i)X_{ip} + \alpha_0^j(u_i, v_i, t_i) + \alpha_1^j(u_i, v_i, t_i)\left(y_{i0}^{j=1} + y_{i0}^{j=2} + y_{i0}^{j=3}\right)$$

$$+ \delta^j(u_i, v_i, t_i)\varepsilon_i^j, \qquad\qquad j = 0, \ldots, J \tag{7.7}$$

$\alpha_0^j(u_i, v_i, t_i)$ can been perceived as the parameter of constant. Let X^*_{ip} denote the vector $[X_{ip}, \ 1, \ \left(y_{i0}^{j=1} + y_{i0}^{j=2} + y_{i0}^{j=3}\right), \ \varepsilon_{it}^j]$. In order to estimate $\beta_p^j(u_i, v_i, t_i)$, $\alpha_0^j(u_i, v_i, t_i)$, $\alpha_1^j(u_i, v_i, t_i)$, and $\delta^j(u_i, v_i, t_i)$ for each point, X^* should be transferred into $\widetilde{X}^*$ in GSTLM, as follows.

$$\widetilde{X}^* = W^{ST}(u_i, v_i, t_i)^{1/2} X^* \tag{7.8}$$

where $W^{ST}(u_i, v_i, t_i)$ is an n by n diagonal matrix:

$$W^{ST}(u_i, v_i, t_i) = \begin{pmatrix} w_{i1}^{ST} & 0 & \cdots & 0 \\ 0 & w_{i2}^{ST} & \cdots & 0 \\ \cdots & \cdots & \ddots & \vdots \\ 0 & 0 & \cdots & w_{i3}^{ST} \end{pmatrix} \tag{7.9}$$

Based on d^{ST}, the elements of $W^{ST}(u_i, v_i, t_i)$, w_{im}^{ST} are defined. The same issue arises in the estimation of GSTLM as GTWLM. On the one hand, the estimation of ST-PLM, involving non-linear optimization and MSL estimation, is computationally intensive. Furthermore, if the GSTLM parameters are to be computed at n regression points, then the above procedures must be repeated n times! Thus, when parameters are estimated for every point, sub-samples are chosen by using a computational ruse that ignores observations with negligible weight to reduce the computational burden. On the other hand, a sub-sample with values of all single types cannot be produced. Hence, adaptive weighting functions are employed to adapt themselves in size to ensure that the same and sufficient non-zero weights are used for each in the analysis. The specific adaptive weighting function is as follows:

$$w_{im}^{ST} = \begin{cases} \exp\left\{-\left(\dfrac{(d_{im}^{ST})^2}{h_{ST}^2}\right)\right\}, & \text{if } m \text{ is the } q \text{ nearest points around } i \\ 0, & \text{otherwise} \end{cases} \tag{7.10}$$

where h_{ST} is same spatio-temporal bandwidth as in Equation 7.5. q represents the number or proportion of observations to consider when estimating the regression at each location to reduce the number of sub-samples and make enough so that the regression can work. Substituting 6.4 in 6.5 and 6.10 results in the following:

$$w_{im}^{ST} = \exp\left\{-\left(\frac{\lambda[(u_i - u_m)^2 + (v_i - v_m)^2] + \mu(t_i - t_m)^2}{h_{ST}^2}\right)\right\}$$

$$= \exp\left\{-\left(\frac{(u_i - u_m)^2 + (v_i - v_m)^2}{h_S^2} + \frac{(t_i - t_m)^2}{h_T^2}\right)\right\}$$

$$= \exp\left\{-\left(\frac{(d_{im}^S)^2}{h_S^2} + \frac{(d_{im}^T)^2}{h_T^2}\right)\right\}$$

$$= \exp\left\{-\frac{(d_{im}^S)^2}{h_S^2}\right\} \times \exp\left\{-\frac{(d_{im}^T)^2}{h_S^T}\right\}$$

$$= w_{im}^S \cdot w_{im}^T \tag{7.11}$$

where $w_{im}^S = \exp\left\{-{(d_{im}^S)^2}\big/{h_S^2}\right\}$, $w_{im}^T = \exp\left\{-{(d_{im}^T)^2}\big/{h_S^T}\right\}$, $(d_{im}^S)^2 = (u_i - u_m)^2 +$ $(v_i - v_m)^2$, $(d_{im}^T)^2 = (t_i - t_m)^2$, $h_S^2 = \frac{h_{ST}^2}{\lambda}$ and $h_T^2 = \frac{h_{ST}^2}{\mu}$ are the spatial and temporal bandwidths, respectively. Because the weighting function is a diagonal matrix, whose diagonal elements are multiplied by $w_{im}^S \cdot w_{im}^T$ $(1 \le m \le n)$, W^{ST} can be seen as the combination of spatially weighted matrix W^S and temporally weighted matrix W^T : $W^{ST} = W^S \times W^T$. Hence, if μ was set to 0, GSTLM becomes a GSLM which considers only the spatial effect. If λ was set to 0, the GSTLM becomes a GTLM which considers only the temporal effect.

In order to reduce the parameters to be estimated in the model, the parameter ratio $\tau = \mu/\lambda$ is introduced. It can be considered as the balance ratio between the spatial distance effect and the temporal distance effect. As discussed in Chapter 4, setting $\lambda = 1$ will not change the results estimated by optimizing λ, μ, and h_{ST} by using cross-validation in terms of PCP. Hence, w_{im}^{ST} can be expressed as follows:

$$w_{im}^{ST} = \exp\left\{-\left(\frac{[(u_i - u_m)^2 + (v_i - v_m)^2] + \tau(t_i - t_m)^2}{h_{ST}^2}\right)\right\} \tag{7.12}$$

Here, the estimation results will provide τ and h_{ST} instead of λ, μ, and h_{ST} in this study. In GSTLM, the CV method is used to choose an appropriate bandwidth h_{ST} and sub-sample size q by maximum PCP. The weighting matrix is decided subsequently. Following that, the parameters in each point can be estimated by the local log-likelihood function, which is formulated as follows:

$$\ln(L(u_i, v_i, t_i))$$

$$= \sum_{i=1}^{N} \sum_{j=0}^{J} y_{it}^j \ln(p(y_i^j = 1 | \widetilde{X}_i^*; \beta_p^j(u_i, v_i, t_i), \alpha_0^j(u_i, v_i, t_i), \alpha_1^j(u_i, v_i, t_i), \delta^j(u_i, v_i, t_i))$$

$$\tag{7.13}$$

Theoretically, the parameters of the local model are estimated by maximizing the log-likelihood function. However, that ε_i^j in $\widetilde{X}_i^*$ is assumed to be random complicated the computation of the probabilities for each land use type. The local log-likelihood function above involves the estimation of an integral, which cannot be evaluated analytically owing to its lack of a closed-form solution. Therefore, MSL is adopted (same as ML, except that the simulated probabilities are used instead of the exact probabilities). Generalized Cholesky decomposition is also employed to decompose the matrix given by Equation 7.5.

The estimations for GSTLM in each point are obtained from y by the following steps.

Step 1) Decompose the covariance matrix into a triangular matrix by the generalized Cholesky decomposition method.

Step 2) Obtain three (for $j = 1$, 2, and 3) sequences of draws following the *iid* normal distribution by shuffled Halton draws. Each sequence consists of n number for each cell.

Step 3) Multiply the lower-triangular matrix by n *iid* normal values and get two new sequences.

Step 4) Calculate the Equation 7.13 in accordance with ε_i^j given by the new sequences.

Step 5) Repeat steps 2 to 4 100 times and average the results.

Step 6) Maximize the simulated log-likelihood function and get parameters.

7.3 Results and Discussion

To check the applicability of GSTLM and to analyze the land use change in SEZ, Shenzhen, a study was implemented by using the spatio-temporal sample set between 1990 and 2008.

The accuracy and parameter estimate results are reported in Tables 7.1 and 7.2, respectively. The PCP of GSTLM reaches a value of 85.9%, which is higher than MNLN, GTWLM, and ST-PLMB. However, the PCP improvement of GSTLM on MNLM is less than the sum of the improvement on MNLM by GTWLM and ST-PLM; (MNLM: 74.1%; GTWLM: 82.3%; ST-PLM: 79.4%). That is because allowing for non-stationarity in the regression parameters can account for at least some, and possibly a large part, of the autocorrelation in error terms in a global model calibrated with spatio-temporal data. Also, the accuracies for all the land use types achieved by GSTLM are balanced. Because the output of local parameter estimates from GSTLM would be voluminous, Table 7.2 provides a three-number summary of the distribution of each parameter to indicate the extent of the variability. In this case, the impact of those variables varies spatially, and it indicates that local effects do exist.

Furthermore, two problems remain to be considered. Firstly, it needs to be ascertained whether it is necessary to consider individual effect and spatio-temporal autocorrelation in the model. Secondly, it needs to be found whether the GSTLM is truly better than MNLM, GTWLM, and ST-PLM.

For the first problem, a pseudo t-statistic is calculated to indicate the significance of individual effect and spatio-temporal autocorrelation. This is obtained by dividing a parameter estimate by its standard error (Fotheringham *et al.*, 2001). If the absolute t-statistic value is greater than 1.96, it can be considered significant at the 95% confidence level. Table 7.3 lists the absolute t-statistics for all the parameters. Figure 7.1 provides the spatial variation of the t-statistic for individual effect and spatio-temporal autocorrelation.

Table 7.1 The accuracy of GSTLM

Observed	Predicted				Total	PCP
	0	1	2	3		
0	1,844	19	7	107	1,977	93.3%
1	31	354	14	147	546	64.8%
2	4	6	255	35	300	85.0%
3	124	77	21	1,155	1,377	83.9%
Total	2,003	456	297	1,444	4,200	85.9%

Table 7.2 GSTLM parameter estimate summaries

| Parameter | GSTLM (bandwidth = 0.6633; τ = 0.1826; q = 1130) | | | | | | | | |
| | Residential | | | Industrial | | | Transportation/commercial/others | | |
	Min	Med	Max	Min	Med	Max	Min	Med	Max
Distance to commercial center	−3.79e−02	−1.13e−03	1.22e−02	−2.62e−02	−2.16e−03	9.95e−03	−1.87e−02	1.86e−03	1.09e−02
Distance to financial center	−1.87e−02	−1.59e−03	1.92e−02	−1.65e−02	6.84e−05	2.69e−02	−5.11e−03	1.70e−04	1.50e−02
Distance to industrial center	−1.85e−02	−5.57e−04	2.34e−02	−3.49	−8.48e−02	−9.09e−03	−1.42e−02	2.96e−04	1.42e−02
Distance to educational facilities	−5.34e−02	−2.55e−03	9.70e−03	−2.50e−02	6.78e−03	6.42e−02	−1.40e−02	−6.95e−04	1.79e−02
Distance to railway infrastructure	−6.41e−03	−6.09e−04	6.99e−03	−2.27e−02	−2.01e−03	1.05e−03	−8.49e−03	−7.38e−04	1.08e−03
Distance to road	−7.01e−02	−1.45e−02	8.94e−03	−6.61e−02	7.26e−03	8.09e−02	−7.73e−02	−1.11e−02	1.14e−02
Population	−9.09e−04	2.17e−04	3.16e−03	−8.47e−04	3.01e−04	7.27e−03	−3.80e−04	4.19e−04	3.24e−03
DEM	−1.12	−1.18e−01	1.90e−02	−8.30e−01	−1.24e−01	1.28e−01	−3.99e−01	−6.11e−02	−2.80e−03
Slope	−3.45	−4.31e−01	5.70e−01	−4.53	−5.56e−01	4.15e−01	−2.64	−2.88e−01	5.45e−01
Planning for residential	−1.14e01	1.10e01	9.60e02	−4.06e01	3.83	6.53e01	−1.10e01	7.16	5.53e01
Planning for industrial	−2.13e02	1.27e02	5.16e02	−3.61e01	1.32e02	6.35e02	−2.53e01	1.34e02	5.46e02
Planning for transportation/commercial/others	−2.91e01	−1.31e−01	3.85e01	−8.78	5.63	3.50e01	−7.44	4.70	2.81e01
No planning	−4.31e02	−5.32e−01	5.83e01	−5.50e02	1.36	5.69e01	−4.09	3.42	3.12e01
α_0^j	−2.71e01	5.56	3.49e01	−3.83e01	13.8e01	6.21e01	−15.7e01	−1.10	1.65e01
α_1^j	−1.71e01	8.55	2.65e02	−2.30e01	4.96	2.69e02	6.19	2.22e01	2.99e02
δ^j	−1.28e02	4.79	2.33e02	−4.84e02	−5.25e01	3.69e02	−1.83e02	1.91e01	5.09e02
$\ell\ell$ =					−2.3898e+003				

Table 7.3 GSTLM absolute t estimate summaries

Parameter	Residential			Industrial			Transportation/commercial/others		
	Min	Med	Max	Min	Med	Max	Min	Med	Max
Distance to commercial center	3.09e-03	1.19	5.68	1.96e-03	8.51e-01	3.22	6.74e-04	1.81	5.38
Distance to financial center	1.12e-03	1.55	4.75	8.53e-04	1.10	4.08	1.38e-03	1.25	5.40
Distance to industrial center	5.25e-04	1.11	4.95	2.05	7.05	10.0	3.37e-04	9.73e-01	4.07
Distance to educational facilities	1.39e-04	8.79e-01	6.09	7.85e-03	1.60	5.33	3.20e-03	1.54	4.10
Distance to railway infrastructure	1.10e-03	1.26	5.32	2.90e-04	2.01	6.99	3.68e-04	1.54	6.86
Distance to road	4.03e-03	2.03	4.71	1.84e-04	1.52	5.14	5.59e-03	3.40	7.27
Population	3.68e-05	1.19	4.53	1.19e-03	1.21	5.73	2.59e-03	2.02	6.54
DEM	1.81e-03	2.30	5.47	1.12e-03	1.16	4.87	3.62e-02	2.52	6.67
Slope	3.94e-03	2.57	6.25	1.59e-03	2.06	4.26	2.30e-03	3.34	7.57
Planning for residential	1.49e-03	3.10	6.80	1.06e-03	1.38	4.63	5.10e-04	2.00	5.94
Planning for industrial	4.65e-07	4.81e-02	8.02e-02	1.72e-04	4.87e-02	9.99e-02	2.93e-07	4.85e-02	7.61e-02
Planning for transportation/commercial/others	7.80e-05	1.07	3.86	1.19e-03	1.25	3.89	7.34e-03	2.58	4.91
No planning	2.34e-04	1.10	4.04	1.28e-03	1.28	3.66	1.28e-05	1.39	5.63
a^i_0	1.01e-03	1.48	4.60	1.92e-04	1.71	5.34	7.05e-05	8.08e-01	4.56
a^i_1	4.10e-04	2.39	5.54	2.07e-03	1.40	4.56	3.50e-02	7.27	9.85
δ^i	1.26e-04	8.23e-01	3.33	3.20e-04	1.37	4.13	4.47e-04	1.04	6.40

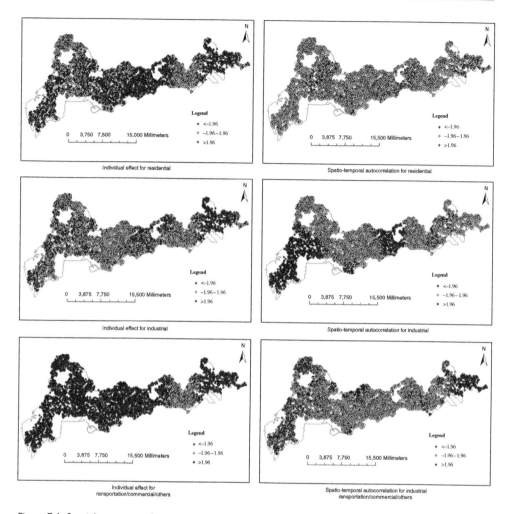

Figure 7.1 Spatial variation of t-statistics for individual effect and spatio-temporal autocorrelation

From Table 7.3 it can be seen that the median value of absolute t-statistic for residential and transportation/commercial/others are 2.39 and 7.27, respectively. This implies that the individual effect is significant for most of the observations for residential and transportation/commercial/others, especially in Luohu and Futian . For the industrial land use type, the individual effect is primarily significant in the north-west of Yantian.

The smaller median value of the absolute t-statistic indicates that only part of the observations have significant spatio-temporal autocorrelation, whereas the others have weak spatio-temporal autocorrelation with the absolute t-statistic values (< 1.96). The median value of absolute t for residential is 0.82 (Table 7.3). This shows that the spatio-temporal autocorrelation of the tested data for residential is rather weak. Besides, the median values of absolute t for the other two land use types are about 1, which evinces that the spatio-temporal autocorrelation for these two land use types is not significant on some observations. One important reason is that only 700 observations were collected in each year and the temporal distribution of land use data is thin. Hence, the sample set is sparsely scattered among the spatio-temporal space. However, it is still reasonable to account for spatio-temporal autocorrelation in some observations where the t-statistic values are greater than 1.96. It can be deduced from Figure 7.1 that significant spatio-temporal autocorrelation exists in the south-eastern part of Nanshan and the eastern part of Luohu for industrial land use type, and in the north-eastern part of Yantian. Furthermore, the spatio-temporal autocorrelation is also significant for transportation/commercial/others.

Additionally, a *t*-statistic is calculated to indicate the significance of the other parameters. As is evident from Table 7.3, planning for industrial is not significant on every observation (highlighted).

With regards to the second question mentioned earlier, McNamara's test is employed to test the significant difference between the MNLM, GTWLM, ST-PLM, and GSTLM, and the results are presented in Table 7.4.

The negative values clearly indicate that the GSTLM model outperforms the MNLM, GTWLM, and ST-PLM. The Z values between MNLM, GTWLM, ST-PLM, and GSTLM are -17.30, -6.88, and -13.84, respectively, thereby indicating that GSTLM substantially outperforms the other models. Also, the Z value between GTWLM and ST-PLM is 4.31, which is more than 1.96. This result demonstrates that a significant difference exists between the GTWLM and ST-PLM at the 95% confidence level. This comparison further demonstrates that the GTWLM outperforms the ST-PLM in terms of model accuracy.

Table 7.5 provides the Kappa coefficients and their comparison with the other models. The Kno shows that the GSTLM achieves a better result than all the other models. Specifically, the GSTLM is the most optimal model for specifying the location and the GTWLM is the best model to specify quantity.

Table 7.4 Significance comparison for different models

Models Comparison	Z values		
	GTWLM	ST-PLM	GSTLM
MNLM	−14.16	−8.36	−17.30
GTWLM	–	4.31	−6.88
ST-PLM	–	–	−13.84

Table 7.5 Comparison of MNLM, GTWLM, ST-PLM, and GSTLM with coefficients

	MNLM	GTWLM	ST-PLM	GSTLM
Kno	0.65	0.76	0.73	0.81
Klocation	0.63	0.73	0.71	0.81
Kquantity	0.90	0.98	0.90	0.93

7.4 Analysis of Spatio-temporal Land Use Distribution Pattern in SEZ, Shenzhen

Unlike the global models (MNLM, ST-PLM), which have unified parameters across space and time, the GSTLM generates a set of spatio-temporal parameter estimates on each land use sample observation. These can be used to analyze spatial and temporal variations of the effects of land use pattern determinants. Based on the sample points with parameter estimates, a set of pictures (Figures 7.2 to 7.4) are generated to reveal the spatial variations of explanatory factors for each land use type with generally regular spatial patterns. Then, several parameter estimates, which have remarkable temporal variation, are selected and spatio-temporal analysis is performed. The spatial and temporal variations of those factors are shown in Figure 7.5.

Figure 7.2 presents the spatial variation of parameter estimates for residential land use. The distance to commercial center has a greater negative effect on residential land use in the eastern part of Luohu and the western part of Yantian. Almost all the positive effects from distance to financial center occur in Luohu. Distance to the educational facilities has a more negative effect in the eastern part of the SEZ. This is logical because the educational facilities are scarce resources in the eastern part and hence they have greater effect. Both the influence of distance to railway infrastructure and road are higher negative values. This signifies that the residential land use is distributed in the areas that are relatively closer to the transportation networks. Especially, the effect of proximity to the transportation networks in Futian is less or positive. Population has positive effect, whereas DEM and slope present a negative influence on the residential land use. However, negative influence of population can be found in some regions in the southern part of Nanshan. The planning for residential has a greater positive influence on the residential land use, which implies that the planning for residential land use plays a very crucial role in the study area.

Figure 7.3 provides the spatial variations of parameter estimates for industrial land use. It can be seen that the distance to the commercial/financial center has an additional negative effect on the industrial land use in the western part of the study area compared to the eastern part. The reason for the positive effects on the industrial land use is that most of the commercial and financial centers are located in the middle of the study area, namely, in Luohu and Futian. As expected, the distance to the industrial center has a negative influence throughout the study area, but more so in the northern part of Nanshan. This is where the biggest industrial park, the Nanshan High Technology Park, is located. With the influence on residential land use, the distance to educational facilities has positive effect in the eastern part and negative effect in the western part. The distance to the railway infrastructure and distance to road chiefly affect the industrial land use negatively. Especially, a greater negative effect is found in the eastern part. The influence from population has almost the same pattern as it has in the residential land use. DEM and slope still have negative influence for industrial land use. The planning for industrial land use has a stronger influence

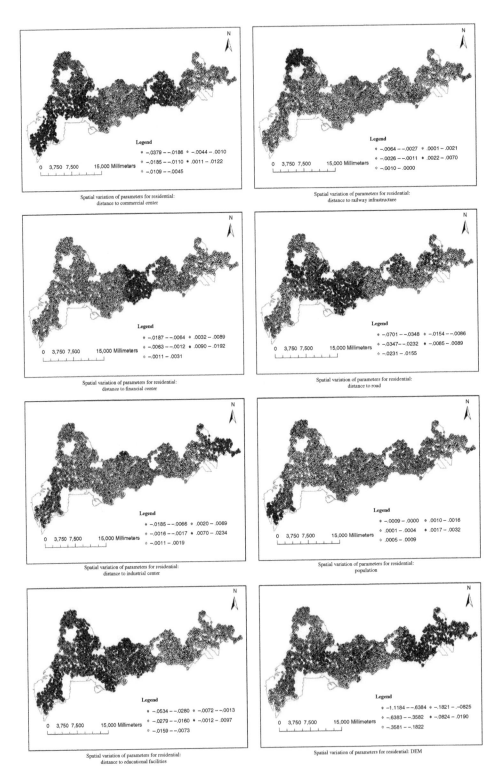

Figure 7.2 Spatial variation of parameters for residential

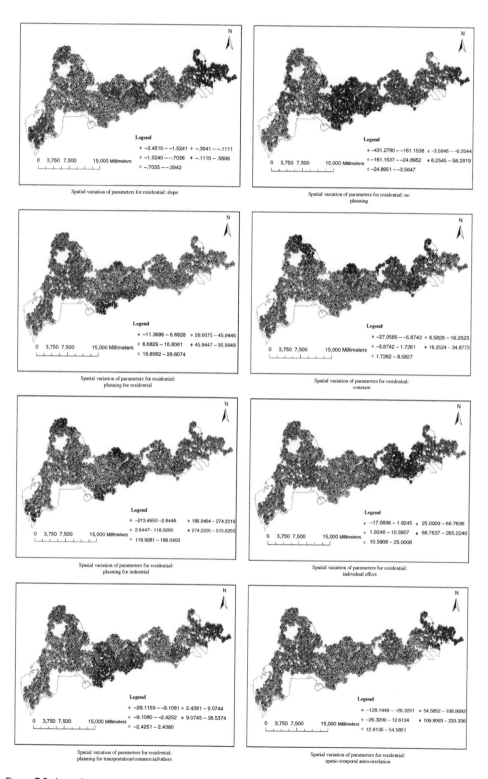

Figure 7.2 (cont.)

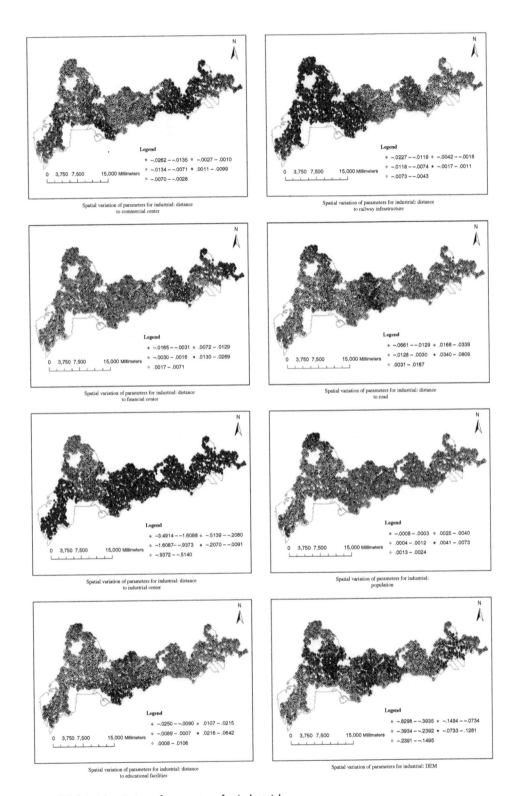

Legend
◦ −.0262 – −.0135　◦ −.0027 – .0010
◦ −.0134 – −.0071　✦ .0011 – .0099
◦ −.0070 – −.0028

Spatial variation of parameters for industrial: distance
to commercial center

Legend
◦ −.0227 – −.0119　◦ −.0042 – −.0018
◦ −.0118 – −.0074　✦ −.0017 – .0011
◦ −.0073 – −.0043

Spatial variation of parameters for industrial: distance
to railway infrastructure

Legend
◦ −.0165 – −.0031　◦ .0072 – .0129
◦ −.0030 – −.0016　✦ .0130 – .0269
◦ .0017 – .0071

Spatial variation of parameters for industrial: distance
to financial center

Legend
◦ −.0661 – −.0129　◦ .0168 – .0339
◦ −.0128 – −.0030　✦ .0340 – .0809
◦ .0031 – .0167

Spatial variation of parameters for industrial: distance
to road

Legend
◦ −3.4914 – −1.6088　◦ −.5139 – −.2080
◦ −1.6087 – −.9373　✦ −.2070 – −.0091
◦ −.9372 – −.5140

Spatial variation of parameters for industrial: distance
to industrial center

Legend
◦ −.0008 – .0003　◦ .0025 – .0040
◦ .0004 – .0012　✦ .0041 – .0073
◦ .0013 – .0024

Spatial variation of parameters for industrial:
population

Legend
◦ −.0250 – −.0090　◦ .0107 – .0215
◦ −.0089 – −.0007　✦ .0216 – .0642
◦ .0008 – .0106

Spatial variation of parameters for industrial: distance
to educational facilities

Legend
✦ −.8298 – −.3935　◦ −.1494 – −.0734
◦ −.3934 – −.2392　✦ −.0733 – .1281
◦ −.2391 – −.1495

Spatial variation of parameters for industrial: DEM

Figure 7.3 Spatial variation of parameters for industrial

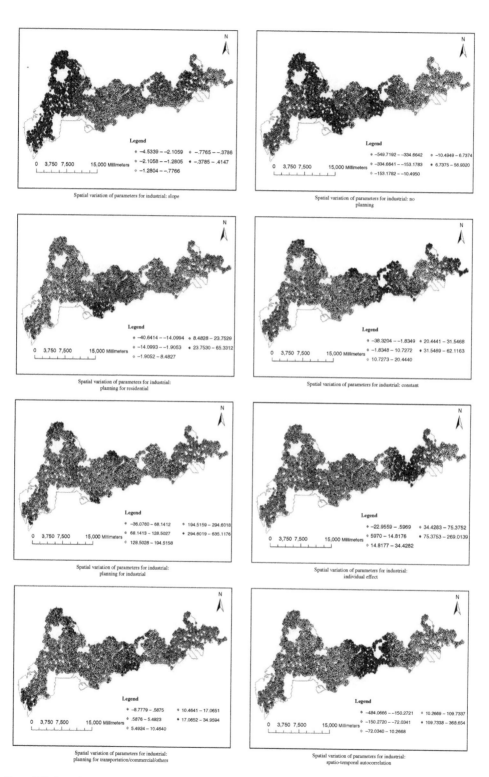

Figure 7.3 (cont.)

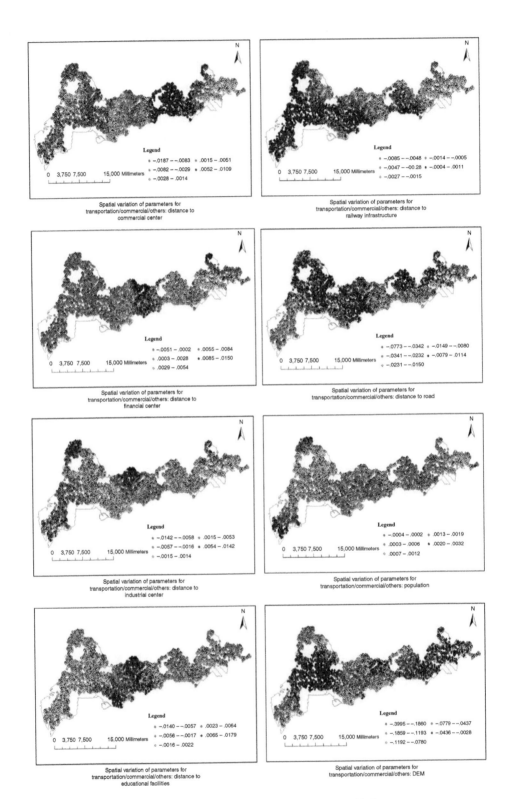

Figure 7.4 Spatial variation of parameters or transportation/commercial/others

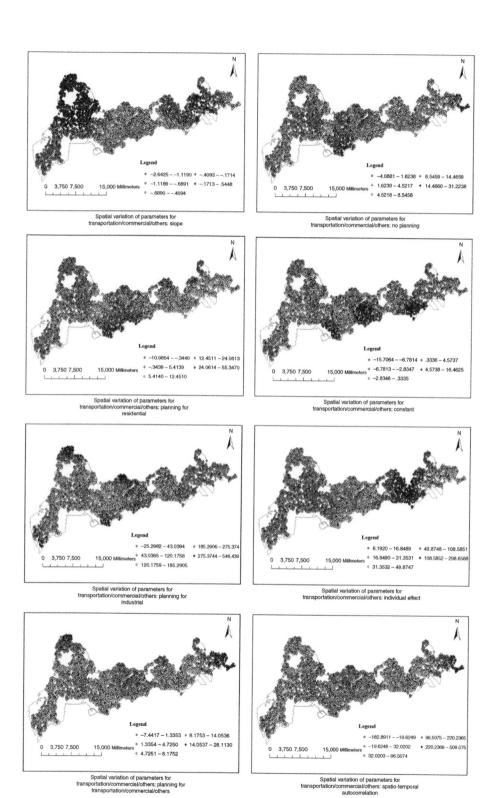

Figure 7.4 (cont.)

with large parameter estimates. The picture also reveals that most of the industrial areas are built in accordance with the planning regulations except in the middle eastern part of Yantian.

Figure 7.4 illustrates that the distance to commercial center has an added negative influence for the transportation/commercial/others land use type. Also, the distance to the financial center and educational facilities have a more positive effect at the junction of Luohu and Futian. Distance to the railway infrastructure and road have negative influence. In particular, it can be seen that the positive influence of the distance to transportation network is mainly concentrated in the western part of the study area. Distance to the industrial center has a negative effect in the study area. However, positive influence of the distance to the industrial center can still be found in some parts. Similar to residential and industrial land use, DEM and slope have negative influence for transportation/commercial/others. For some part, population has a negative effect on transportation/commercial/others in the study area. Planning for transportation/commercial/others performs well except in the old district, Luohu.

Table 7.6 shows that the median parameter estimates change over time. It can be found that some parameter estimates have a remarkable change in different years. In order to analyze the spatio-temporal relationship between the land use pattern and the explanatory factors, only those explanatory factors which show changes on sign are selected. Figure 7.5 illustrates the spatio-temporal variations

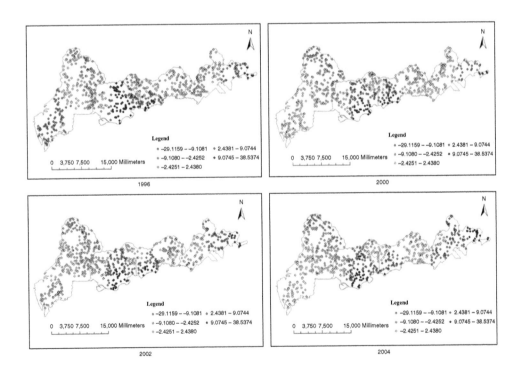

Figure 7.5 Spatio-temporal variation of parameters

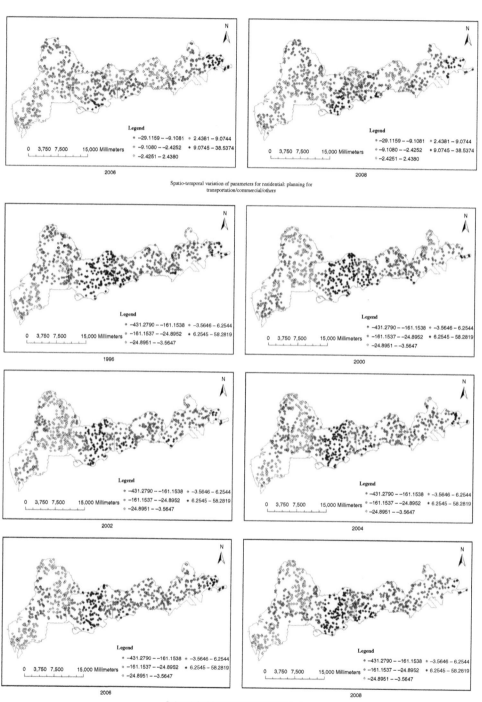

Spatio-temporal variation of parameters for residential: planning for transportation/commercial/others

Spatio-temporal variation of parameters for residential: no planning

Figure 7.5 (cont.)

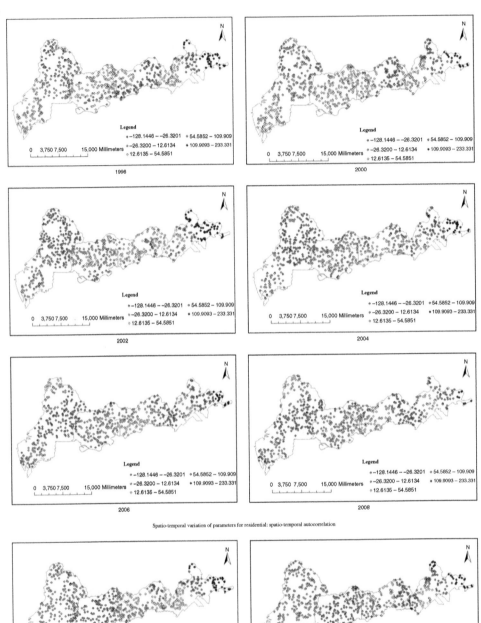

Spatio-temporal variation of parameters for residential: spatio-temporal autocorrelation

Figure 7.5 (cont.)

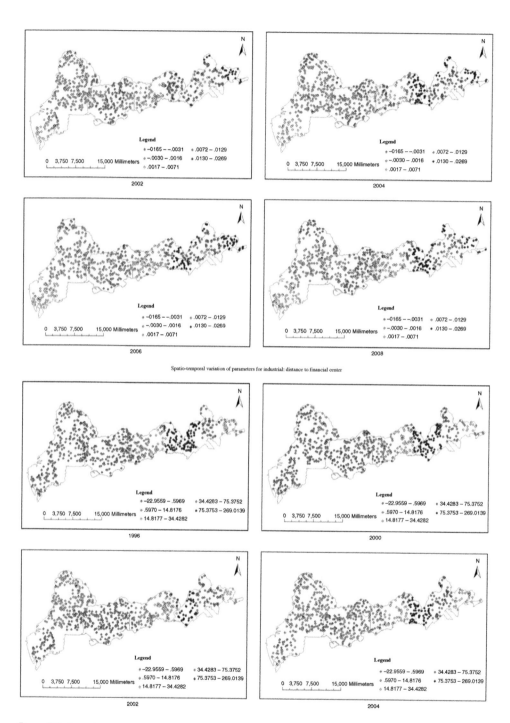

Spatio-temporal variation of parameters for industrial: distance to financial center

Figure 7.5 (cont.)

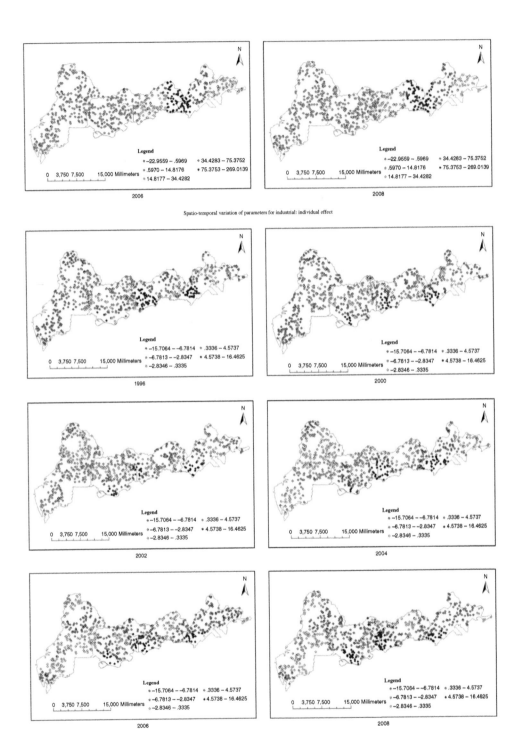

Figure 7.5 (cont.)

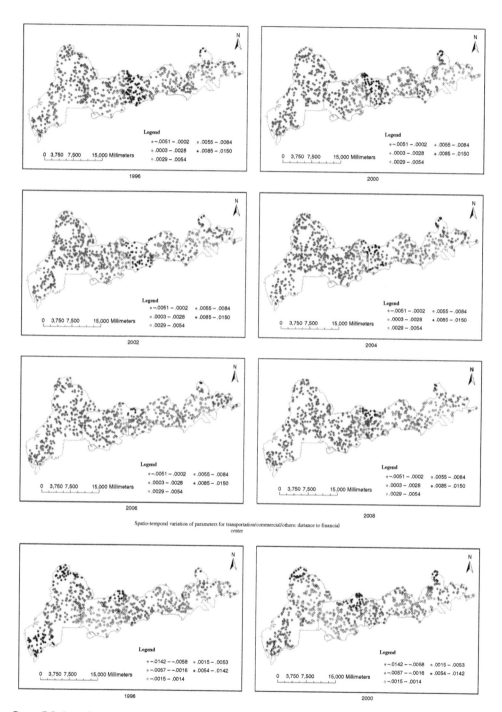

Spatio-temporal variation of parameters for transportation/commercial/others: distance to financial center

Figure 7.5 (cont.)

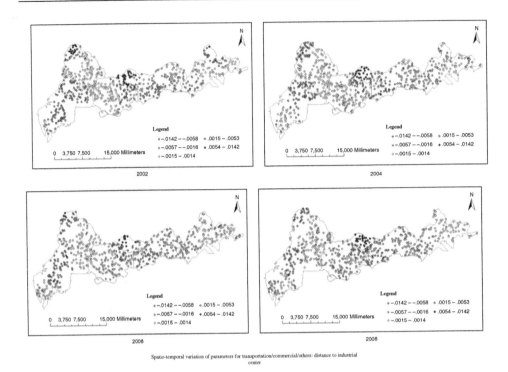

Spatio-temporal variation of parameters for transportation/commercial/others: distance to industrial center

Figure 7.5 (cont.)

of the selected parameter estimates over different years. The parameter, distance to the financial center for industrial land use, is considered, for example. In the SEZ, high technology industry is the major component of industry. It is a kind of capital-intensive industry, and it largely depends on the financial market. Hence, the distribution of financial center has a strong relationship with industry land use. During the early years of city development, most industryl land use is built near the financial center. For Shenzhen, almost all the financial organizations were located in Futian at the beginning. Thus, the parameter estimates in Futian are negative over time. As time goes by, the land near the financial center becomes more expensive. Meanwhile, with the development of finance, the financial organizations can cover the whole SEZ. The new industry land use depends less on the distance to the financial center located in Futian. More and more industry land use can be built on in the eastern part of Luohu and the western part of Yantian, which are undeveloped. As seen from the pictures, the colors of the parameter estimates gradually change from yellow to red between 1996 and 2008 in the eastern part of Luohu and the western part of Yantian. This implies that the distance between the financial center and the industrial land use is greater in this part.

Table 7.6 GSTLM median estimate in different years

	Residential					
Time	1996	2000	2002	2004	2006	2008
Distance to commercial center	−1.65e−03	−3.70e−04	−4.09e−04	−1.51e−03	−1.34e−03	−1.13e−03
Distance to financial center	−1.31e−03	−1.94e−03	−1.59e−03	−1.66e−03	−1.62e−03	−1.44e−03
Distance to industrial center	−1.85e−05	−5.03e−04	−1.04e−03	−1.45e−04	−9.71e−05	−1.38e−03
Distance to educational facilities	−1.81e−03	−2.33e−03	−3.38e−03	−2.72e−03	−2.40e−03	−1.60e−03
Distance to railway infrastructure	−4.67e−04	−4.94e−04	−6.65e−04	−7.11e−04	−6.20e−04	−6.24e−04
Distance to road	−1.09e−02	−1.37e−02	−1.43e−02	−1.56e−02	−1.59e−02	−1.69e−02
Population	3.31e−04	2.10e−04	1.08e−04	1.53e−04	1.76e−04	2.16e−04
DEM	−1.54e−01	−1.30e−01	−1.21e−01	−1.12e−01	−9.71e−02	−9.78e−02
Slope	−4.34e−01	−4.52e−01	−4.85e−01	−4.47e−01	−3.93e−01	−4.23e−01
Planning for residential	1.17e+01	1.12e+01	1.02e+01	1.05e+01	1.06e+01	1.14e+01
Planning for industrial	1.59e+02	1.19e+02	1.18e+02	1.08e+02	1.27e+02	1.36e+02
Planning for transportation/ commercial/others	−3.89e−01	−8.33e−01	−5.33e−01	3.35e−01	8.97e−01	1.07
No planning	8.53e−01	−2.57	−1.30	−2.64e−01	−6.30e−01	2.59e−01
α_0^j	4.52	5.19	6.45	6.42	5.63	5.02
α_1^j	4.37	4.22	4.98	8.78	1.19e+01	1.32e+01
δ^j	3.53e+01	2.19e+01	2.94	−8.85	−1.31e+01	−1.16e+01

	Industrial					
Time	1996	2000	2002	2004	2006	2008
Distance to commercial center	−1.97e−03	−1.24e−03	−2.45e−03	−2.71e−03	−2.16e−03	−2.39e−03
Distance to financial center	−9.80e−05	−5.57e−04	−3.36e−04	2.04e−04	1.32e−03	1.76e−04
Distance to industrial center	−8.84e−02	−7.39e−02	−1.01e−01	−9.78e−02	−7.93e−02	−8.24e−02
Distance to educational facilities	6.77e−03	7.19e−03	8.09e−03	7.69e−03	4.97e−03	5.67e−03
Distance to railway infrastructure	−2.39e−03	−2.47e−03	−1.80e−03	−1.59e−03	−2.05e−03	−1.64e−03
Distance to road	7.81e−03	5.51e−03	7.05e−03	8.44e−03	6.86e−03	7.67e−03
Population	7.64e−04	4.25e−04	4.20e−04	2.40e−04	2.24e−04	2.57e−04

Table 7.6 (cont.)

	Industrial					
Time	1996	2000	2002	2004	2006	2008
DEM	−1.48e−01	−1.39e−01	−1.45e−01	−1.36e−01	−1.00e−01	−1.04e−01
Slope	−7.72e−01	−5.55e−01	−4.79e−01	−5.28e−01	−5.40e−01	−5.56e−01
Planning for residential	6.44	5.58	3.02	2.89	3.04	3.02
Planning for industrial	1.46e+02	1.36e+02	1.30e+02	1.28e+02	1.23e+02	1.37e+02
Planning for transportation/ commercial/others	1.03e+01	7.91	4.87	3.64	3.94	5.16
No planning	3.63	7.32e−01	1.03	6.72e−01	5.28e−01	1.74
α_0^j	1.21e+01	9.38	1.35e+01	1.56e+01	1.40e+01	1.35e+01
α_1^j	−2.96	−7.25e−01	1.83	3.91	7.34	9.33
δ^j	−8.45e+01	−6.5 8e01	−4.89e+01	−4.31e+01	−4.22e+01	−3.55e+01

	Transportation/commercial/others					
Time	1996	2000	2002	2004	2006	2008
Distance to commercial center	1.09e−03	1.75e−03	2.28e−03	2.08e−03	1.73e−03	1.60e−03
Distance to financial center	6.08e−04	−1.25e−05	1.01e−04	−2.15e−04	−7.67e−05	2.07e−04
Distance to industrial center	1.01e−03	1.29e−03	7.70e−04	−2.24e−04	−4.77e−04	−1.26e−03
Distance to educational facilities	−9.86e−04	−1.20e−03	−4.74e−04	−5.29e−04	−7.86e−04	−3.59e−04
Distance to railway infrastructure	−4.80e−04	−7.07e−04	−7.62e−04	−8.16e−04	−8.07e−04	−7.91e−04
Distance to road	−6.44e−03	−7.05e−03	−8.22e−03	−1.24e−02	−1.61e−02	−1.76e−02
Population	6.27e−04	5.00e−04	4.37e−04	3.41e−04	2.94e−04	3.38e−04
DEM	−8.22e−02	−7.14e−02	−6.98e−02	−5.14e−02	−4.48e−02	−4.64e−02
Slope	−2.87e−01	−2.76e−01	−2.80e−01	−3.06e−01	−2.86e−01	−3.07e−01
Planning for residential	5.32	7.23	7.51	8.53	7.51	7.12
Planning for industrial	1.61e+02	1.39e+02	1.36e+02	1.22e+02	1.24e+02	1.29e+02
Planning for transportation/ commercial/others	4.76	5.37	4.87	5.09	4.02	3.50
No planning	3.88	4.31	3.91	3.06	2.40	2.47
α_0^j	−4.27	−3.85	−3.13	−7.77e−01	8.44e−01	9.84e−01
α_1^j	2.37e+01	2.08e+01	1.92e+01	2.10e+01	2.35e+01	2.41e+01
δ^j	2.83e+01	4.18e+01	2.82e+01	1.40e+01	7.94e+00	1.18e+01

Part III

Land Use Multi-objective Optimization

Chapter 8

Optimization Objectives Setting

This chapter firstly analyzes the main environmental problems experienced by Shenzhen during its rapid development stage; these relate to its soil, water, and ecosystem. To solve these problems in the planning process, several objectives, which are in accord with the basic principles, are defined at different levels of the municipal overall land use plan and the urban master land use plan. The objectives of the former focus on the problems of the entire city, whereas those of the latter concentrate on developments in urban areas rather than in municipal areas. These two parts of the objectives are presented in detail in Sections 8.3 and 8.4, respectively.

8.1 Environmental Problems

The rapid development of Shenzhen in the last three decades has resulted in severe environmental problems, such as soil erosion, nonpoint source (NPS) pollution, and excessive carbon emissions. This publication attempts to define and measure these environmental problems and to evaluate the relationships of these issues with spatial land use change. Then, based on the quantified relationships, an optimized spatial plan will be implemented.

Soil erosion

Shenzhen is located in a mountainous area which has steep slopes (Chen *et al.*, 2012), where land elevation above 80 meters occupies approximately 30% of the total area, and with the lowest elevation is at sea level. Shenzhen is located in a subtropical monsoon climate zone and therefore it has abundant rainfall. Because soil erosion is often the process of soil being washed away by the flow of water, Shenzhen's steep slopes and abundant rainfall aggravate the process of soil erosion (Lee, 2004; Dumas *et al.*, 2010). Therefore, Shenzhen is naturally vulnerable to severe soil erosion.

In addition to this natural vulnerability, a significant amount of land use change aggravates soil erosion in Shenzhen. Some scholarship has confirmed the severe soil erosion generated during the construction process in Shenzhen (Li & Zhang, 2002; Mu *et al.*, 2010).

Water pollution

Based on the municipal overall land use plan of Shenzhen (2006–2020), the water resource per capita in Shenzhen is approximately 600m^3, which is approximately one-

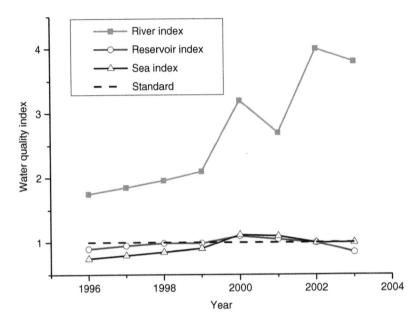

Figure 8.1 Water quality of Shenzhen from 1996 to 2003
Source: Zhang (2007)

third of the total national average, and a quarter of the total of Guangdong Province. The water resource data indicate that Shenzhen is one of the cities suffering from a severe water shortage (Chen *et al.*, 2011). Water pollution is the main cause of this water resource scarcity. Based on the water quality index (Zhang, 2007), shown in Figure 8.1, the reservoir index and sea index were approximately similar to the standard value from 1996 to 2003, while the river index was higher than the standard value, which indicated that the river water quality in Shenzhen had worsened.

Excessive carbon emissions

Along with its increasing economic development and population growth, China has become one of the world's leading CO_2 emitters (Yang *et al.*, 2012). It contributed 13.5% of global CO_2 emissions in 2000, which made it the world's second largest emitter after the United States of America (Zhang, 2000). Furthermore, China's contribution is expected to exceed that of the United States of America by 2020 if the country continues to develop at its current rate (Zhang, 2000). Therefore, the study of CO_2 emissions in China is significant for CO_2-emissions reduction on the global scale. The problem of excessive CO_2 emissions is severe in China's cities, especially those undergoing rapid development, such as Beijing, Shanghai, Guangzhou, and Shenzhen (Dhakal, 2009; Zhang & Cheng, 2009; Li *et al.*, 2010). Because rapidly developing cities in China are suffering the consequences of excessive CO_2 emissions at first hand, their experiences are critical to the future of other cities in the country. Shenzhen will be China's first low-carbon ecological demonstration city, and it is intended by the Ministry of Housing and Urban–Rural Development to provide a good example to the whole country. In addition, it has measurable heat islands and elevated carbon levels (Chen *et al.*, 2012; Xie *et al.*, 2013).

Landslides

Since the 1980s, landslides have been a perennial problem in Shenzhen, causing a significant loss of life and property (Tian *et al.*, 2008). As mentioned previously, Shenzhen features numerous mountainous areas with steep slopes (Chen *et al.*, 2012). Rapid land use changes likewise aggravate landslide occurrence in Shenzhen. Land use change is recognized throughout the world as one of the most important factors influencing the occurrence of rainfall-triggered landslides (Glade, 2003).

Ecosystem service value

Shenzhen is one of the fastest-growing metropolitan areas in China. As a result, urban sprawl in Shenzhen significantly affects the value of ecosystem services and functions. Some academic studies on the ecosystem service value in Shenzhen have been conducted. Li *et al.* (2010) observed a 231.3 million RMB decrease in ecosystem service value from 1996 to 2004 in Shenzhen, a trend mainly caused by the decreasing areas of forest, wetland, and water body. Besides cities, rural areas are also facing these problems (Peng *et al.*, 2017). Within this context, maximizing ecosystem service value is selected as one objective.

8.2 Basic Principles

Based on the background of land use problems during the development of Shenzhen, and referring to national and local policies as well as land use principles, a series of objective principles is proposed for the spatial planning of land use in Shenzhen.

Sustainable land use objectives

The objectives are selected based on the principle of sustainable land use. Sustainable land use focuses on the balance between the environment, economy, and society. Therefore, the objective system should involve these three aspects at the same time. Specifically, the environmental objectives should include water quality, carbon dioxide control, reserve protection, and development restriction based on the slope and digital elevation model. The economic objectives should include the area of built-up land in the projected year and the targeted GDP in the projected year. The social objectives should include the change cost, compatibility of land use, and spatial equality.

Smart growth land use objectives

For smart growth, the objectives based on mixed land use and accessibility should be considered.

New urbanism land use objectives

For new urbanism, in addition to the objectives of smart growth, the objectives based on various housing and high-density areas should be considered.

Land use objectives in the Chinese context

In addition to the aforementioned objectives that are based on the land use planning principles and the characteristics of Shenzhen, some specific land use planning objectives in the Chinese context are also proposed.

The first constraint is the constraint on the area of basic farmland. China needs to protect and preserve its farmlands as a key point because of the significant demand for food from a large population. Therefore, in the spatial planning of land use for Shenzhen, the constraint on the area of basic farmland to protect farmlands should be met first.

The second constraint is the targeted GDP in the projected year. Given that the 12th Five-Year Plan for National Economic and Social Development of Shenzhen listed in detail the expected GDP increase rate for the succeeding years, the spatial planning of land use should correspond to that requirement.

At the same time, some Chinese laws or regulations set specific requirements on land use planning. For example, the rule on built-up land per capita for different cities, stated in the code for the classification of urban land use and planning standards for land development (Ministry of Housing and Urban-Rural Construction of the People's Republic of China, 2011), requires that these requirements should be met by the spatial planning of land use.

All of these constraints or objectives should be considered in land use planning in China. Therefore, in the spatial planning of land use in Shenzhen, all of these constraints or objectives are considered.

8.3 Objectives at the Municipal Overall Land Use Plan Level

Spatial planning of land use is conducted at two different levels. At the municipal overall land use plan level, nine objectives and four constraints were proposed. The nine objectives are: maximizing economic benefit; maximizing ecosystem service value; minimizing soil erosion; minimizing NPS pollution; minimizing carbon emissions; maximizing compatibility; minimizing change cost; maximizing accessibility; and minimizing landslide susceptibility. Constraints are also set on the areas of built-up land, water body, and cultivated land as well as on reserve protection.

The objectives of maximizing mixed land use and maximizing accessibility are determined based on smart growth and new urbanism. For sustainable land use, the balance between three aspects, namely, environment, economy, and society, should be considered and reflected in the multiple objective system. Soil erosion, carbon emissions, landslide, and ecosystem service value are also considered based on the background of Shenzhen. For example, the objective of minimizing soil erosion is selected because Shenzhen is located in a mountainous area with steep slopes (Chen *et al.*, 2012) in a subtropical monsoon climate zone with abundant rainfall. These features make

Shenzhen naturally vulnerable to severe soil erosion. With regard to the constraints on built-up land, basic farmlands are considered on the basis of local or national policies.

These nine objectives are conflicting. For example, maximizing economic benefit requires built-up land, whereas maximizing ecosystem service value requires forest. Similarly, compatibility makes the attempt to reallocate land use in space to ensure a highly compatible land use pattern, whereas minimizing change cost attempts to reduce change. The conflict comes from the fact that the objectives are proposed by different stakeholders from different fields. Limited land use resources aggravate the conflict.

8.4 Objectives at the Urban Master Plan Level

The objectives proposed here are the goals of the spatial planning of land use that should be achieved at the urban master plan level. These objectives are detailed and involve land use allocation within the urban area, which has different objectives at the municipal overall land use plan level.

In urban areas, firstly, housing capacity should be sufficient to accommodate the population; secondly, employment capacity should be sufficient to provide citizens with jobs. These two objectives are the basic functions of cities. In order to maintain good living conditions in urban areas, the objectives of maximizing green space and minimizing pollutants from industrial land are considered as planning objectives in this publication. Socially related objectives, such as accessibility and spatial equity, are considered. Moreover, spatial objectives, which control the spatial layout of urban land use, such as maximizing compatibility between land use zones and minimizing change cost, are considered.

Based on the urban land use structure of the urban master plan of Shenzhen (2010–2020) (Shenzhen Municipal Government, 2010), constraints on the ratios of residential, administrative, and public services as well as on industrial and green spaces are also proposed.

Similarly, the objectives at the urban master plan level are conflicting. For example, the objective of maximizing housing capacity is in conflict with the objective of maximizing employment capacity. Mixing land use patterns will increase the cost of land use changes. The pattern of mixed land use is different from that of compatible land use. The conflict comes from the different requirements of stakeholders and the limitation of land use resources.

Chapter 9

Optimization at the Overall Land Use Plan Level

In this chapter, the detailed definitions and measurements of objectives at the overall land use plan level are proposed firstly. They focus on the problems of the entire city. According to basic principles, the proposed objectives include the maximization of economic benefit, ecosystem services value, compatibility and accessibility, and the minimization of soil erosion, NPS pollution, carbon emission, change cost, and landslide susceptibility. As for constraints followed by objectives, they are built-up land area, water body, cultivated land, elevation, and reserve protection.

9.1 Objectives at the Overall Land Use Plan Level

9.1.1 To Maximize the Economic Benefit

The unit GDP factor, Un_GDP, for a given land use, which indicates the GDP generated in the unit area for the given land use (RMB/m^2), varies from one land usage to another and from one year to another. Historical Un_GDP can be retrieved from the statistical GDP value and the area of the corresponding land usage.

The primary GDP in Shenzhen comprises the planting, forestry, animal husbandry, and fisheries, each of which is generated on different land usage. Both secondary and tertiary industries in Shenzhen are generated on built-up land.

With regard to primary GDP, planting comprises grain crops, peanuts, vegetables, and fruits. The GDP of fruits is generated on garden plot, whereas the other components of planting GDP are generated on cultivated land. With regard to animal husbandry, the main livestock in Shenzhen are pigs, cattle, and chickens. Therefore, animal husbandry GDP in Shenzhen is generated on cultivated land. In terms of fishery, both marine and freshwater fish are found in Shenzhen. Freshwater fish thrive in the water bodies in Shenzhen; marine fish thrive in the sea, which is not included in this publication. Table 9.1 lists the grouped GDP and the corresponding land use.

The historical Un_GDP for each land use type can be calculated (see Table 9.2) based on the reparation of the GDP value and the corresponding area of each land use. The temporal variations in Un_GDPs on cultivated land, garden plot, forest, and water body are negligible. Land use efficiency generally improves over time. Therefore, the maximal Un_GDPs on these land use types are used in the projected year.

However, the temporal variation of built-up land should be considered. According to the Shenzhen Economic and Social Development Twelfth Five-Year Plan (Shenzhen Municipal Government, 2011), the area of built-up land that generates 1,000 RMB is

Table 9.1 GDP grouped by sector and corresponding land uses

	GDP grouped by sector		Corresponding land uses
The primary GDP	Planting	Grain crops	Cultivated land
		Peanuts	Cultivated land
		Vegetable	Cultivated land
		Fruit	Garden plot
	Forestry		Forest
	Animal husbandry	Pigs, cattle and chickens	Cultivated land
	Fishery	Freshwater-fish	Water body
		Marine-fish fishery	–
The secondary and tertiary industries	–	–	Built-up land

Table 9.2 GDP value generated on per unit area of different land use types

Year	Cultivated land	Garden plot	Forest	Water body	Built-up land
	RMB/m^2				
1996	29.83	0.51	0.04	1.06	237.68
2000	28.58	0.40	0.09	0.47	444.12
2002	39.23	0.53	0.07	0.50	516.50
2004	39.53	0.51	0.10	0.37	664.10
2006	23.74	0.15	0.06	0.27	857.80
2008	35.38	0.08	0.09	0.13	1,031.04
2020	39.53	0.53	0.10	1.06	2,339.72

expected to decrease by 25% from 2010 to 2015. Meanwhile, the annual decease rate of economic development in Shenzhen is 5.59% from 2010 to 2015. In 2008, the area of built-up land that generates 10,000 RMB was 9.70m^2. With that rate, the $9.7*(1 - 0.0559)^7 = 6.48m^2$ built-up land will generate 10,000 RMB in 2015. Assuming that the decease rate from 2015 to 2020 will be slightly larger than that from 2010 to 2015, it can conclude that the area of built-up land that generates 10,000 RMB in 2020 will be $6.48 \times (1 - 0.08)^5 = 4.27m^2$, and that the GDP value generated on per m^2 of built-up land is 2,339.72 RMB.

According to the preceding calculation, the Un_GDP on cultivated land is set to 39.53 RMB/m^2, that on garden plot to 0.53 RMB/m^2, that on forest to 0.1 RMB/m^2, that on water body to 1.06 RMB/m^2, and that on built-up land to 2,339.72 RMB/m^2. Subsequently, the objective can be quantified through Equation 9.1.

$$max \; Z_{GDP} = \sum_{n=1}^{8} Un\_GDP_n * Area_n \qquad (9.1)$$

where Un_GDP$_n$ is the GDP value generated on per unit area of the n-th land use type, and Area$_n$ is the area of the n-th land use type.

9.1.2 To Maximize Ecosystem Services Value

The economic valuation of ecosystem services has been widely used to understand the multiple benefits provided by ecosystems (Zhao *et al.*, 2004; Chuai *et al.*, 2016). Ecosystem services can be defined as conditions and processes through which natural ecosystems and species sustain and realize human life (Li *et al.*, 2010). Landscapes are the bases of all terrestrial ecosystems. Changes in land use inevitably alter the types, areas, and spatial distributions of various ecosystems (Luo & Zhang, 2013). Based on this context, numerous studies have focused on ecosystem service evaluation from the perspective of land use change (Zhao *et al.*, 2004; Wang *et al.*, 2006; Quétier *et al.*, 2007; Li *et al.*, 2010; Tianhong *et al.*, 2010). Considering that land use changes influence ecosystem values, some scholars have called for a sound land use plan to maintain a high ecosystem value. However, only a few studies have exerted efforts to formulate a land use plan that considers ecosystem values. In this publication, ecosystem service value is defined as one of the objectives in the spatial planning of land use, as follows:

$$max\ Z_{ESV} = \sum_{n=1}^{8} Area_n \times VC_n \tag{9.2}$$

where Z_{ESV} is the objective of ecosystem service value, $Area_n$ is the area of the n-th land usage, and VC_n is the coefficient value (RMB/ha/yr) for the n-th land use type.

Costanza *et al.* (1998) divided the global biosphere into 16 types of ecosystem and 17 types of service function. Costanza *et al.* (1998) estimated the ecosystem service value for each type of ecosystem. For the ecosystem service value used in China, Xie *et al.* (2003) extracted the equivalent weight factor for ecosystem services per hectare of terrestrial ecosystems by surveying 200 Chinese ecologists and different regions across China and localizing average natural food production. Referring to the work conducted for China and Shenzhen (Zhao *et al.*, 2004; Wang *et al.*, 2006; Li *et al.*, 2010; Luo & Zhang, 2013), VC_n on different land usages can be determined through existing studies in Shenzhen (see Table 9.3).

Table 9.3 Ecosystem service value of unit area of different land usages in Shenzhen (RMB/ha/yr)

Land use type	VC
Cultivated land	6,689.5
Garden plot	14,080.9
Forest	21,152.8
Grassland	7,009.0
Built-up land	0
Transportation	0
Water body	44,542
Unused land	406.6

9.1.3 To Minimize Soil Erosion

In this part, the well-know USLE proposed by Wischmeier (1976) is employed to calculate the soil erosion with regard to land use changes. The USLE has since become one of the most widely used empirically grounded approaches (Ozcan *et al.*, 2008; Meusburger *et al.*, 2010; Abu Hammad, 2011). It consists of a set of calculations to estimate soil erosion on a plot of land with homogeneous characteristics (Wischmeier & Smith, 1978). The USLE model allows the average annual soil loss based on the product of five erosion risk indicators (Meusburger *et al.*, 2010). In the USLE model, the average annual soil loss is based on the product of five erosion risk indicators (see Equation 9.3):

$$minZ_{soil\ erosion} = A \times R \times K \times LS \times C \times P \tag{9.3}$$

where $Z_{soil\ erosion}$ (ton ha^{-1} a^{-1}) is the predicted average annual soil loss, R (MJ mm $^{ha^{-1}\ h^{-1}}$ year^{-1}) is the rainfall and runoff factor, K (ton ha^{-1}per unit R) is the soil erosivity factor, LS (dimensionless) is the slope length and slope steepness factor, C (dimensionless) is the cover and management factor, and P (dimensionless) is the conservation practices factor. The R, K, and LS factors basically determine the erosion volume, whereas the C and P factors are reduction factors ranging from 0 to 1 (Kumar & Kushwaha, 2013). The P factor reflects the impact of land use changes, with one land use type maintaining a unique P factor. When the land use changes, the quantitative area of each land use type changes, and the rainfall, slope, soil type, and vegetation cover for that type also change. The USLE model thus allows the influence of land use change on soil erosion to be reflected.

Meanwhile, considerable research has confirmed the occurrence of severe soil erosion from construction sites during rapid and extensive urbanization. Farnworth *et al.* (1979) and Lemly (1982) agreed that urban development contributes to soil erosion acceleration. Soil erosion rates on construction sites are two to 4,000 times greater than during preconstruction conditions; moreover, they are important components of NPS pollution (Harbor, 1999). As the amount of built-up land in the optimal land use plan increases, the soil erosion rate during construction rises. Therefore, both land use pattern and land use change in the optimized plan affect soil erosion. Based on this information, Equation 9.4 should be revised as follows:

$$minZ_{soil\ erosion} = A \times R \times K \times LS \times C \times P + CA_{k5} * Agg \tag{9.4}$$

where CA_{k5} is the area of the LUZ that changes from the k-th land use type in the current land use pattern to built-up land in the optimal plan. Agg is the soil loss rates from construction sites. Table 9.4 lists the reviewed Agg values in different countries that were proposed in various years. According to the values listed in Table 9.4, the Agg in Shenzhen was 611 tons/ha/year and 584 tons/ha/year, as proposed by different studies. Using the Agg values in other countries as references, it deems that the maximum soil loss rate is typically smaller than or approximate to 500 tons/ha/year. Based on this context, the Agg in Shenzhen is determined as 584 tons/ha/year.

Table 9.4 Literature review on soil loss rates for construction sites.

Case study area	Land use	Characteristic	Soil loss rate (tons/ha/year)	Reference
South Korea	Athletic facilities	Max	210	Maniquiz et al. (2009)
Maryland	Building sites	Max	550	Wolman and Schick (1967)
Illinois	Typical construction site	Max	490	Maniquiz et al. (2009)
Georgia	Road	Max	500	Maniquiz et al. (2009)
Oklahoma	Typical construction site	Max	1,230	Maniquiz et al. (2009)
Shenzhen	Typical construction site	Average	611	Liu (1995)
Shenzhen	Mechanical plantation ground	Average	584	He et al. (1993)
Dongguan	Quarrying	Average	797.68	Wang et al. (2006)
Guangzhou	Typical construction site	Average	658.42	Wang et al. (2006)
Huiyang	Laying the high-tension line	Average	567.84	Wang et al. (2006)

Rainfall and runoff factor: R

The rainfall and runoff factor (R) represents the two characteristics of a storm that determine its erosivity: the amount of rainfall, and the peak intensity sustained over an extended period. Drawing on Li et al. (2009), R can be calculated by the following equation:

$$R = \sum_{i=1}^{12}(0.3046Pi - 2.6398) \tag{9.5}$$

where R is the rainfall erosivity factor (MJ mm ha^{-1} h^{-1}yr^{-1}) and Pi is the monthly average rainfall (mm) (Kim et al., 2005). Table 9.5 lists the annual rainfall (Shenzhen Statistics Bureau, 1997, 2001, 2003, 2005, 2007, 2009) in 1996, 2000, 2002, 2004, 2006, and 2008 and corresponding values of R factors for Shenzhen in various years.

The value of the R factor for Shenzhen can be obtained. However, this R factor is constant for the whole of Shenzhen and does not reflect spatial differences. Therefore, the spatial distribution of the annual average rainfall for numerous years is used to spatialize it (see Equation 9.6). More specifically, this spatial distribution is achieved by

Table 9.5 Annual precipitations of Shenzhen from 1996 to 2008

Year	Annual rainfall [mm]	R factor [MJ mm ha^{-1} h^{-1}yr^{-1}]
1996	1,683.3	481.06
2000	2,533.6	740.06
2002	1,882.8	541.82
2004	1,299.4	364.12
2006	1,936.5	558.18
2008	2,710	793.79

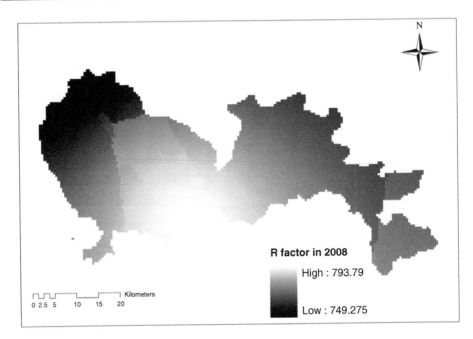

Figure 9.1 Spatial distribution of R factor in 2008 Soil erosivity factor: K

interpolation, as the annual average rainfall for numerous years is available at each monitoring station.

$$R_{(m,n)} = \frac{AnAvRainfall_{(m,n)} - min(AnAvRainfall)}{max\ (AnAvRainfall) - min(AnAvRainfall)} R \qquad (9.6)$$

where $R_{(m,n)}$ is the value of refined R factor at the location of grid (m,n), AnAvRainfall$_{(m,n)}$ is the value of the annual average rainfall for numerous years at grid (m,n), min(AnAvRainfall), and max(AnAvRainfall) are the minimum and maximum of the annual average rainfall for numerous years in the entire city. Finally, with the assistance of the GIS tool, the R factor with spatial information is achieved. Figure 9.1 shows the spatial distribution of the R factor in 2008 as an example.

The soil erosivity factor is defined as the rate of soil loss per unit of R as measured on a unit plot (Ozsoy *et al.*, 2012), and it represents the average long-term soil and soil profile response to the erosive power associated with rainfall and runoff (Lee, 2004). The K factor is determined by the soil's physical and chemical properties, which vary from place to place. Several experimental models have been established on the basis of soil texture, organic matter, structure, and osmosis (Wischmeier, 1976; Wischmeier & Smith, 1978). In this dissertation, the value of K is determined by the following equation given by Sharpley and Williams (1990):

$$K = \{0.2 + 0.3\exp[-0.0256w_d(1 - w_t/100)]\}$$
$$\times \left(\frac{w_i}{w_i + w_l}\right)^{0.3} \times \left\{1.0 - \frac{0.025w_c}{w_c + \exp(3.72 - 2.95w_c)}\right\} \quad (9.7)$$
$$\times \left\{1.0 - \frac{0.7w_n}{w_n + \exp(-5.51 + 2.29w_n)}\right\}$$

where, w_d is the percentage of sand, w_i is the percentage of silt, w_l is the percentage of clay, w_c is the percentage of organic matter, and it can be obtained by the formula: $w_n = (1 - \frac{w_d}{100})$ (Sharpley & Williams, 1990; Zhou et $al.$, 2008). All these soil characteristics data used in Equation 9.7 are available from the Food and Agriculture Organization of the United Nations (Freddy Nachtergaele et $al.$, 2008). The spatial distribution of the K factor for Shenzhen is obtained by using Equation 9.7 and the GIS tool, and it is presented in Figure 9.2.

The slope length and slope steepness factor (LS) represents the effect of the topography on soil erosion (Lufafa et $al.$, 2003); this is because an increase in slope length and steepness produces higher overland flow velocities and therefore stronger erosion (Dumas et $al.$, 2010). LS is derived from Equation 9.8, as follows:

$$LS = \left(\frac{\lambda}{22.13}\right)^m \times (65.41sin^2\theta + 4.56sin\theta + 0.0065) \quad (9.8)$$

where λ is the slope length in meters, θ is the slope angle in degrees, and m is a slope angle contingent variable (McCool et $al.$, 1987) that can be calculated by Equation 9.9.

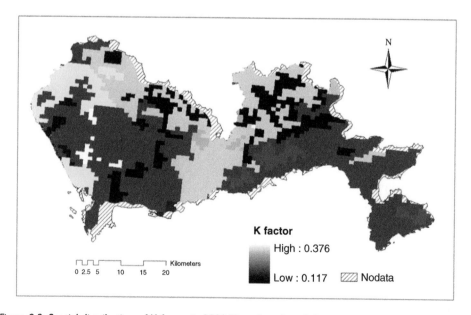

Figure 9.2 Spatial distribution of K factor in 2008 Slope length and slope steepness factor: LS

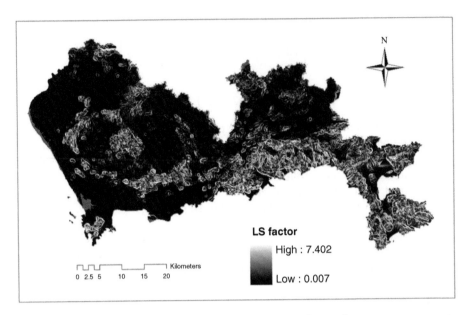

Figure 9.3 Spatial distribution of LS factor Crop and management factor: *C*

$$m = \begin{cases} 0.3 & 22.5^0 \le \theta \\ 0.25 & 17.5^0 \le \theta < 22.5^0 \\ 0.2 & 12.5^0 \le \theta < 17.5^0 \\ 0.15 & 7.5^0 \le \theta < 12.5^0 \\ 0.10 & \theta < 7.5^0 \end{cases} \tag{9.9}$$

The coefficients of λ and θ are obtained from the digital elevation model, the resolution of which is 30m by 30m. Finally, based on these equations, the spatial distribution of the LS factor for Shenzhen is obtained, and it is represented in Figure 9.3.

The crop and management factor C depends on vegetation cover, which dissipates the kinetic energy of raindrops before they hit the soil surface. Erosion and runoff are markedly affected by different types of vegetation cover. Erosion and runoff are measured as the ratio of soil loss to land cropped under continuously fallow conditions (Wischmeier & Smith, 1978). According to this definition, C equals 1 when it is subject to standard fallow conditions. When the percentage of vegetation cover approaches 100%, the C factor value approaches the minimum. The C factor can be calculated by Equation 9.10, as follows (Cai *et al.*, 2000; Ma *et al.*, 2001; Zhao *et al.*, 2007) and lc by the normalized differential vegetation index (NDVI) (see Equation 9.11):

$$C = \begin{cases} 1 & lc \le 10\% \\ 0.6805 - 0.3436 lg\ lc & 10\% < lc < 78.3\% \\ 0 & 78.3\% \le lc \end{cases} \tag{9.10}$$

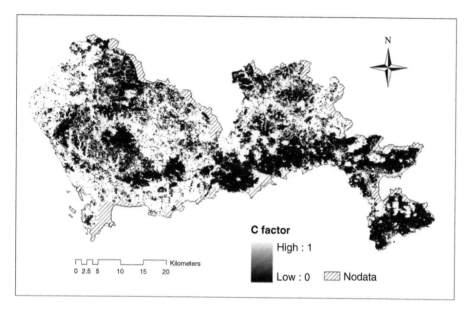

Figure 9.4 Spatial distribution of C factor Conservation practices factor: *P*

$$
lc = \begin{cases} 0 & -1 \leq NDVI \leq -0.0675 \\ \dfrac{NDVI + 0.0675}{0.47} & -0.0675 < NDVI \leq 0.4025 \\ 1 & 0.4025 < NDVI \leq 1 \end{cases} \qquad (9.11)
$$

where lc (dimensionless) is vegetation coverage, and NDVI (dimensionless), which ranges from −1 to 1, can be obtained from the RS image. An NDVI value approaching 1 indicates that the land is fully covered by vegetation, leading to a high value for lc. Using the foregoing equations, the spatial distribution of the C factor for Shenzhen is obtained, as shown in Figure 9.4.

The conservation practice factor P is defined as the ratio of soil loss from the upward and downward slope of an inclined plane where a soil preservation policy has been put in place (Park *et al.*, 2011). In fact, the P factor affects erosion by redirecting runoff around the slope to produce less erosivity or slowing down the runoff to make a deposition (Lee, 2004). Its value ranges from 0 to 1, with a lower value suggesting more effective conservation practices. According to the literature (Xu & Li, 1999; Yang *et al.*, 2003; Zhao *et al.*, 2007; Mu *et al.*, 2010; Park *et al.*, 2011; Kumar & Kushwaha, 2013), the P factor is associated with the land use type and is a reflection of land use changes. The value of the P factor for each type of land use in Shenzhen is determined (Table 9.6) with reference to the literatures.

9.1.4 To Minimize Nonpoint Source Pollution

The objective to minimize the NPS pollution is carried out by the ECM, one empirical model which has been regarded as reliable for simulating NPS pollution (Worrall & Burt, 1999). The ECM approach was originally developed in North America to predict

Table 9.6 P factors of Shenzhen for different land uses types

Land use type	P factor
Cultivated land	0.5
Garden plot	0.6
Forest	1.0
Grassland	1.0
Built-up land	0.3
Transportation	0.3
Water body	0
Unused land	1

nutrient inputs to lakes and streams (Dillon & Kirchner, 1975: Beaulac & Reckhow, 1982), and the process of ECM can be formulated as is shown in Equation 9.12:

$$min\ Z_{NPS} = \sum_{k=1}^{n} A_k E_k \tag{9.12}$$

where Z_{NPS} is the annual pollutant load (tons); E_k is the export coefficient from land use K; A_k is the area occupied by land use K (km^2); and n is the total number of land use classes. The export coefficient expresses the rates at which nitrogen or phosphorus is exported from each source land use area in Shenzhen. It is conventionally derived from literature sources and field experiments (Johnes, 1996). Since the ECM is well developed and has been used widely, the export coefficients are well determined on different land usages. Relative export coefficients of all land use types in Shenzhen are determined by referring to existing literatures (Johnes, 1996; Mattikalli & Richards, 1996; Young et al., 1996; Winter & Duthie, 2000; Li & Zhuang, 2003; Khadam & Kaluarachchi, 2006; Zobrist & Reichert, 2006; Shrestha et al., 2008; Liu et al., 2009) and they are presented in Table 9.7. The ECM model makes it possible to estimate the total annual loads of phosphorus and nitrogen to a water body from NPSs (Winter & Duthie, 2000).

Table 9.7 Export coefficients of the pollutant for different land usages

Land use type	The concentration of dissolved N [tons/km^2/yr]	The concentration of dissolved p [tons/km^2/yr]
Cultivated land	3	0.05
Garden plot	2.2	0.01
Forest	2	0.01
Grassland	1.5	0.02
Built-up land	1.1	0.05
Transportation	1.1	0.05
Water body	0	0
Unused land	0.7	0.051

Source: Winter & Duthie (2000); Liu et al. (2009); Khadam & Kaluarachchi (2006); Mattikalli & Richards (1996); Zobrist & Reichert (2006); Li & Zhuang (2003); Shrestha et al. (2008); Young et al. (1996); Johnes (1996)

9.1.5 To Minimize Carbon Emissions

Before setting the objective of minimizing carbon emissions in Shenzhen, the carbon emissions in Shenzhen have to be evaluated and then the impacts of land use changes on carbon emission growth can be identified. In this part, the theory of carbon sink and carbon source is used to analyze the carbon emission characteristics of Shenzhen. According to the United Nations Framework Convention on Climate Change, carbon sinks are processes, activities, or mechanisms that remove CO_2 from the atmosphere (Yang *et al.*, 2012). On the other hand, a carbon source is any process or activity that releases CO_2 into the atmosphere.

Carbon sinks include forest carbon sinks, grassland carbon sinks, and crop-yield carbon sinks; and the carbon source comprises the carbon source generated from the whole city. In this dissertation, the coefficient approach is used to quantify the carbon sinks, the carbon sources, and the net carbon source.

Calculation of carbon sinks

The main carbon sinks in Shenzhen are forest carbon sinks, grassland carbon sinks, and crop-yield carbon sinks.

Forest carbon sink

The carbon sink constituted by Shenzhen's forests is measured by estimating the forest area and the carbon-sink factor of forest land:

$$Carbon\_Sink_{forest} = C_{forest} \times S_{forest} \qquad (9.13)$$

Carbon_Sink$_{forest}$ is the forest sink (tons), C_{forest} is the carbon-sink factor of the forest (tons C/km^2), and S_{forest} is the area of the forest (km^2). According to existing studies on forest carbon sinks (Fang *et al.*, 2007), together with regional land use characteristics, the carbon-sink factor of Shenzhen's forest land is 4,500 tons C/km^2.

Grassland carbon sink

Similarly, the size of the carbon sink attributable to grassland in Shenzhen is determined by the area of grassland and its carbon-sink factor (Equation 9.14). Based on an existing study (Jiang *et al.*, 2010) and Shenzhen's current land use pattern, the carbon-sink factor of grassland in Shenzhen is 100 tons C/km^2.

$$Carbon\_Sink_{pasture} = C_{pasture} \times S_{pasture} \qquad (9.14)$$

Carbon_Sink$_{pasture}$ is the grassland sink (tons), $C_{pasture}$ is the carbon-sink factor of the grassland (tons C/km^2), and $S_{pasture}$ is the area of the grassland (km^2).

Crop yield carbon sink

Shenzhen's crop sinks are generated by the farming of the grains, peanuts, vegetables and fruit which are the region's major crop types. The quantity of crop sinks in Shenzhen is calculated as follows:

Table 9.8 Water content (r), carbon storage coefficient (C_{crop}), and economic coefficient (P) of the major crops in Shenzhen

Coefficient	Grains	Peanuts	Vegetables	Fruit
r	0.15	0.15	0.75	0.15
C_{crop}	0.4	0.4	0.33	0.3
P	0.4	0.3	0.9	0.1

$$Carbon\_Sink_{crop} = \frac{(1-r) \times Q \times C_{crop}}{P} \tag{9.15}$$

$Carbon\_Sink_{crop}$ denotes the carbon stored in a given crop (tons); r is the water content of the crop, which is dimensionless; Q is the economic yield of the crop (tons); C_{crop} is the carbon-storage coefficient of the crop; and P is the economic coefficient (Yang et al., 2012), which is dimensionless. C_{crop}, P, and r are adopted (Table 9.8) in accordance with the empirical parameters proposed by Li et al. (2004) and with reference to the characteristics of land use in Shenzhen. Data on the economic yield, Q, of each of the four crops, are acquired from the Shenzhen Statistical Yearbooks (Shenzhen Statistics Bureau, 2001; Shenzhen Statistics Bureau, 2003; Shenzhen Statistics Bureau, 2009).

Calculation of carbon sources

The primary carbon source discussed in this publication is the carbon generated by the energy consumption of the whole society. The energy consumption comprises the industrial energy consumption, agricultural energy consumption, traffic energy consumption, and other energy consumptions in Shenzhen. And the carbon source from energy consumption is calculated as follows:

$$Carbon\_Source_{e\_c} = CEF \times E_n \tag{9.16}$$

$Carbon\_Source_{f\_f}$ is the volume of CO_2 emitted by energy consumption (tons), CEF is the CO_2-emissions factor (tons C/ton), and E_n is the amount of energy consumed (tons). The overall quantity of the consumed energy is drawn from the Shenzhen Statistical Yearbooks which is measured by the amount of coal, and the CO_2-emissions factor of coal is obtained from previous studies which is set as 0.97691 (Mika et al., 2010; Carbon Trust, 2011).

Objective defining

In addition to carbon sinks and carbon sources, which reflect the emission and absorption of CO_2 in different cases, estimations of the net carbon source express the amount of CO_2 emissions directly. The objective of reducing carbon emission in Shenzhen can be represented as is shown in Equation 9.17:

$$min \; Z_{Net\_Carbon} = Carbon\_Source - Carbon\_Sink \tag{9.17}$$

Table 9.9 Yield of main crops (Q) in Shenzhen from 1996 to 2008

Year	Grains	Peanuts	Vegetables	Fruit
1996	6,586	735	193,966	36,961
2000	3,700	556	210,479	27,498
2002	3,094	384	217,185	35,279
2004	2,028	98	162,788	28,028
2006	59	6	111,509	12,793
2008	110	4	114,086	4,520

where Z_{Net_Carbon} denotes the net carbon source (tons), Carbon_Source is the total carbon source (tons), and Carbon_Sink is the total carbon sink (tons).

Even if almost all carbon sinks and carbon sources can be calculated based on the area of a certain land use type, for example the carbon sink from forest and grassland, the carbon source generated by energy consumption and the carbon sink of crop yield are determined by the consumed energy and the crop yield which cannot be predicted based on the future land use structure.

The value of the carbon sink of crop yield from 1996 to 2008 is calculated based on the yields of main crops in Shenzhen from 1996 to 2008 (see Table 9.9) which is retrieved from the Shenzhen Statistical Yearbooks (Shenzhen Statistics Bureau, 2009), and the results are listed in the following table.

Therefore, the carbon sink on per unit cultivated land and garden plot are obtained (see Table 9.10) where the carbon sink on per unit cultivated land = (carbon sink of grains, peanuts and vegetables)/the area of cultivated land, and the carbon sink on per unit garden plot = (carbon sink of fruit)/the area of garden plot.

Since the crop yield is not the major economic development element in Shenzhen, the changes in agricultural production system will be insignificant. Hence, the average value of carbon sink generated on per unit cultivated land and garden plot will be used in 2020 which is 330.23 ton, per sq.km and 237.59 ton, per sq.km, respectively.

As for the carbon source from energy consumption, it is found that the energy consumption of per unit GDP, EC_GDP, reduced from 2005 to 2012 (see Table 9.11).

Table 9.10 Carbon sink on per unit cultivated land and garden plot (tons/sq.km)

Carbon sink				Area		Carbon sink on per unit cultivated land	Carbon sink on per unit garden plot
Grains	Peanuts	Vegetable	Fruit	Cultivated land	Garden plot		
5,598.10	833.00	17,780.22	94,250.55	67.42	214.62	359.11	439.15
3,145.00	630.13	19,293.91	70,119.90	74.48	272.30	309.73	257.51
2,629.90	435.20	19,908.63	89,961.45	59.78	301.61	384.30	298.27
1,723.80	111.07	14,922.23	71,471.40	46.93	277.80	357.07	257.28
50.15	6.80	10,221.66	32,622.15	42.49	264.08	241.91	123.53
93.50	4.53	10,457.88	11,526.00	32.06	231.46	329.26	49.80

Table 9.11 Energy consumption of per unit GDP in Shenzhen from 2005 to 2012

Year	Energy consumption per unit of GDP (tons of coal/10,000 yuan)
2005	0.593
2006	0.576
2007	0.56
2008	0.544
2009	0.529
2010	0.513
2011	0.472
2012	0.451

Table 9.12 Prediction model of Energy consumption per unit of GDP

Linear model Poly1: f(x) = p1*x + p2
Coefficients (with 95% confidence bounds): p1 = −0.01988 (−0.02324, −0.01652) p2 = 40.46 (33.71, 47.21)
Goodness-of-fit: SSE: 0.0004749 R-square: 0.9722 Adjusted R-square: 0.9676 RMSE: 0.008897

By the regression model (see Table 9.12), the EC_GDP value will reduce to 0.302 in 2020. Therefore, the carbon source from energy consumption is calculated as follows:

$$Carbon\_Source_{e\_c} = CEF \times GDP_{2020} * EC\_GDP \qquad (9.18)$$

where GDP_{2020} is the GDP value of Shenzhen in 2020.

9.1.6 To Maximize Compatibility

In reality, a given land use, such as an industrial zone, trends to a cluster or a zone which can be defined as compact or contiguous. This phenomenon is used here to minimize the conflicts in neighboring land usage (Ligmann-Zielinska *et al.*, 2008). In addition, each land use type has its own neighborhood preference to choose the neighborhood (Cao *et al.*, 2012). Therefore, it can define the compatibility index between every two land use types. This index ranges from 0 to 1, with a high value indicating a high compatibility between these two land usages. The compatibility index between two land usages that is equal to 1 indicates that these two land usages are totally compatible.

Table 9.13 Compatibility index between land use types

Land use type	LCL	LGP	LF	LGL	LBL	LT	LWB	LUL
LCL	1	0.8	0.9	0.9	0.2	0.5	1	0.5
LGP	0.8	1	0.9	0.8	0.3	0.5	1	0.5
LF	0.9	0.9	1	0.9	0.1	0.1	1	0.3
LGL	0.9	0.8	0.9	1	0.2	0.2	1	0.5
LBL	0.2	0.3	0.1	0.2	1	1	0.3	0.8
LT	0.5	0.5	0.1	0.2	1	1	0.2	0.8
LWB	1	1	1	1	0.3	0.2	1	0.2
LUL	0.5	0.5	0.3	0.5	0.8	0.8	0.2	1

Usually, same land use type may have compatibility index as 1. It means that if two land use zones with the same land use type, they are compatible. On the other hand, the compatibility index between two land use types that is equal to 0 indicates that it is totally incompatible when allocating these two land use close to each other. For example, the incompatibility index between built-up land and water body may be low, since in the real world the built-up land tends to be close to built-up land, while water body prefers to allocate close to forest, grassland or water body. In the process of setting the compatibility index, the opinions of experts, decision-makers, and the involved stakeholders of land use allocation have to be taken into account. To be specific, firstly, the concept of compatibility between land use zones is introduced to the planners, experts, and decision-makers. Secondly, the analytic hierarchy process (AHP) is used to extract the opinions. The AHP is a pair-wise comparison approach (Saaty, 1980), where the compatibility index is compared in pair. The final conflicting degrees are presented in Table 9.13.

The compatibility of each plan can be indicated by adding all the compatibility indexes between each plot and its neighborhood. In addition, the larger area of a given LUZ will lead to a higher conflict with its neighbor. Therefore, the summed product of the compatibility indexes and the areas of the given LUZ and the neighbor is acted as the indicator to reflect the compatibility. The larger the indicator, the more compatible the alternative plan will be. Then, the compatibility of the proposed plan can be formulated by Equation 9.19:

$$\max Z_{com} = \sum_{i=1}^{LUZ_i \in U_{urban}} \sum_{j=1}^{LUZ_j \in NB_i} ComInd_{k,m} \times Aera_i \times Aera_j \times w_{i,j} \qquad (9.19)$$

where Z_{Com} is compatibility of a certain plan; NB_i is the neighborhood of the i-th LUZ; $Area_i$ and $Area_j$ are the area of i-th LUZ and j-th LUZ; $ComInd_{k,m}$ is the compatibility index between the k-th land use type and the m-th land use type; and $w_{i,j}$ is spatial weight between i-th LUZ and j-th LUZ.

Determining the neighborhood of each LUZ is difficult when the compatibility of the entire land use pattern is calculated. In raster data, land use is organized in a matrix. Thus, identifying the neighborhood of a center grid is easy because all grids are represented as a square with the same size. Meanwhile, land uses in vector data are organized in an unstructured shape. Zones that share a border are typically defined as a

Figure 9.5 Spatial location of LUZs

neighbor, which is a binary relationship wherein 1 means neighbor and 0 means not neighbor. However, similar to the situation in Figure 9.5, even if both LUZs a and b are neighbors of LUZ c and the sizes of the three LUZs are the same, the effects of LUZs a and b on LUZ c still differ. Based on this context, the spatial weights matrix is used to index the neighborhood of each LUZ. The spatial weights matrix is an N × N matrix. One row and one column are set for each LUZ. The cell value for any given row/column combination is the weight that quantifies the spatial relationship between the row and column LUZs. Spatial weights that range from 0 to 1 suggest the relationship between two given LUZs. For example, if the spatial weight between LUZs a and b is 0, then they are not neighbors; if the spatial weight is 0.5, then they are neighbors but their effect on each other is 0.5.

By using the spatial weights matrix, it is easy to find out. Determining the neighbors of each urban (ULUZ) is easy when the spatial weights matrix is used. Moreover, the binary relationship between the LUZs is extended into a variable relationship, which is more accurate than the former. The spatial weights matrix can be generated by via ArcGIS software: ArcToolbox/Spatial Statistics Tools/Modeling Spatial Relationships/ Generate Spatial Weights Matrix; and the process that shows how the toolbox works to calculate the spatial weights matrix is released is provided in following link.

http://resources.arcgis.com/en/help/main/10.1/index.html#/How_Generate_Spatial_Weights_Matrix_works/005p00000034000000/

9.1.7 To Minimize Change Cost

Changing cost between two land usages indicates the cost of changing from a certain kind land usage to other one. To compute the changing cost for a certain plan, we need to sum over the changing costs of all changed LUZs. The changing cost can be measured in terms of the population which is affected by the redevelopment, referring to Balling *et al.*'s 2004 study, and it also can be measured in the finance that the change incurs. In this publication, a changing cost factor of each land use is defined to reflect the unit cost of each land use. Because it is difficult to measure the affected population or the financial cost of the change, in this publication a changing cost factor without unit which ranges from 0 to 1 is used. And a higher value indicates higher cost in finance. By doing so, it is simple for the specialists and professionals to represent their opinions on the changing cost factors. The AHP is used here to extract the interviewers' opinions. Firstly, the changing cost factors in existing studies (Chazan *et al.*, 2001; Stewart *et al.*, 2004) are introduced to the interviewers. Then, the interviewers are asked to express their opinions by comparing the elements in pair. The results are listed in Table 9.14. Then, the objective of minimizing the changing cost of a plan is represented, as Equation 9.20 shows:

Table 9.14 Changing cost from one land usage to another

Change from \ Change to	LCL	LGP	LF	LGL	LBL	LT	LWB	LUL
LCL	0	0.3	0.3	03	0.7	0.7	0.1	0.5
LGP	0.4	0	0.2	0.2	0.7	0.6	0.1	0.3
LF	0.4	0.2	0	0.2	0.7	0.6	0.1	0.3
LGL	0.4	0.2	0.2	0	0.7	0.6	0.1	0.3
LBL	1	1	1	1	0	1	1	1
LT	1	1	1	1	1	0	1	1
LWB	1	1	1	1	0.7	1	0	1
LUL	0.2	0.2	0.2	0.2	0.1	0.2	0.5	0

$$min\ Z_{change} = \sum_{j=1}^{N} change\_cost_{k,m} \times Area_j \tag{9.20}$$

where $change\_cost_{k,m}$ is the changing cost from k-th land use type to m-th land use type (see Table 9.14); $Area_j$ is the area of j-th LUZ, and N is the number of LUZs.

9.1.8 To Maximize Accessibility

Accessibility is important for land use planning because it reflects the operational efficiency of a city and boosts social equity (Lindsey et al., 2001; SmoyerTomic et al., 2004; Cao et al., 2011).

Accessibility involves physical distances and time, as well as social, cultural, and gender-based constraints (SmoyerTomic et al., 2004). However, accessibility in our analysis involves only the physical distance. Based on this context, the minimal distance to the nearest road is set as the index to measure the accessibility of LUZs, $Access_{LUZ_i}$. The accessibility of a given land use plan can be measured as follows:

$$max\ Z_{accessibility} = \sum_{m}^{i=1} Access_{LUZ_i}\ (LUZ_i\ is\ built-up\ land) \tag{9.21}$$

In particular, Shenzhen has four road classes that demonstrate different effects on accessibility. Based on the Regulations for Gradation and Classification on Urban Land (GB/T18507-2001) (Ministry of Land and Land Use Management Division China et al., 2001), the effects of road networks on the land use value should be exponential decay. Therefore, the effect of a given class of road on a given LUZ can be measured by the following equation:

$$Access_{LUZ_i, road_k} = \begin{cases} I_k^{1-\frac{dis}{R}} & dis \le R \\ 0 & dis > R \end{cases} \tag{9.22}$$

Where $dis_{i,k}$ is the nearest distance of the i-th LUZ to the k-th class of road, I_k is the influential index of the k-th class of road to the i-th LUZ. R is the influence distance of road. According to the Regulations for Gradation and Classification on Urban Land

Table 9.15 Influence indexes of different classes of road in Shenzhen

Class of road	Influence index	Influent distance (km)
The first class	100	4.5
The second class	80	2
The third class	65	0.75
The fourth class	50	0.35

(GB/T18507-2001) (Ministry of Land and Land Use Management Division China *et al.*, 2001), the influential distances for first-class and second-class roads are determined by the length of the road and the total area of the built-up land. In addition, the influential distances for third-class and fourth-class roads should range between 0.3 km and 0.75 km. The influence index of each road class is listed in Table 9.15 according to the literature review and the Regulations for Gradation and Classification on Urban Land (GB/T18507-2001) (Ministry of Land and Land Use Management Division China *et al.*, 2001).

$$Access_{plot_i} = max\left(Access_{LUZ_i, road_k}\right), k = 1, 2, 3, 4 \tag{9.23}$$

Figure 9.6 and Figure 9.7 show the road networks in Shenzhen and the distances to different classes of road, respectively.

9.1.9 To Minimize Landslide Susceptibility

Even if numerous studies have focused on landslide susceptibility evaluation with the knowledge that land use changes significantly affect landslides, only a few studies have

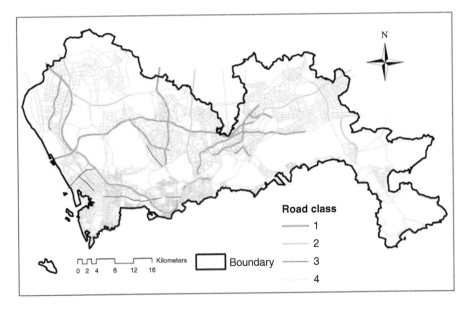

Figure 9.6 Classes of road in Shenzhen

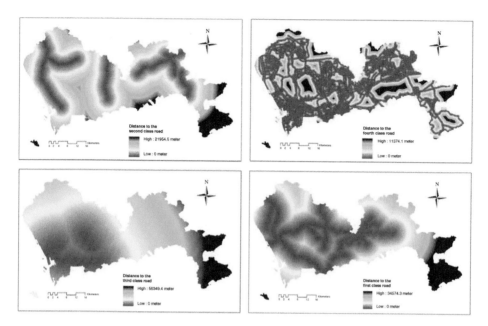

Figure 9.7 Distances to different classes of road in Shenzhen

considered the effect of landslides during land use planning because this phenomenon is complex and difficult to measure. This publication aims to minimize landslide susceptibility during land use planning. To measure landslide susceptibility, a GIS is used to collect and construct landslide-related data. Based on existing studies (Saha *et al.*, 2002; Lee & Talib, 2005; Tian *et al.*, 2008) as well as on the real characteristics and distributions of influential factors, the factors are classified and weighted in Table 9.16, and the corresponding spatial classified data layers are presented in Figure 9.8.

The thrust records are provided by the Data Sharing Infrastructure of Earth System Science, China (www.geodata.cn). The records are non-spatial data that describe the locations of all thrusts in Shenzhen. Therefore, a GIS is used to extend thrust records into space, as is shown in Figure 9.8e.

Based on the existing studies, the landslide hazard index in Shenzhen can be calculated as follows:

$$\mathrm{LHI} = \sum_{i=1}^{7} \mathrm{Weighting}_i * \mathrm{Rate}_i \qquad (9.24)$$

where i denotes the i-th factor that affects the landslide hazard index, $\mathrm{Weighting}_i$ is the weighting number listed in Table 9.16 of the i-th factor, and Rate_i is the rating listed in Table 9.16 and is determined by the value of i-th factor.

Therefore, the objective of reducing landslide damage is measured as follows:

$$\min Z_{\mathrm{landslide}} = \sum_k \mathrm{Area}_k * \mathrm{LHI}_k \qquad (9.25)$$

Table 9.16 Various data layers and landslide hazard weighting-rating system adopted in this study

Data Layers/attributions	Classes	Weighting	Rating
Thrust (buffer)	A. < 500m	9	9
	B. 500–1,000m		6
	C. > 1,000m		3
Land use/land cover	A. Barren	7	9
	B. Sparse vegetation		8
	C. Built-up area		5
	D. Agriculture		4
	E. Forest		3
	F. Water body		1
Slope angle	A. > 45°	5	9
	B. 36–45°		7
	C. 26–35°		5
	D. 16–25°		3
	E. ≤ 15°		1
Relative relief	A. > 120m	4	9
	B. 91–120m		7
	C. 61–90m		5
	D. 31–60m		3
	E. ≤ 30m		1
Human activity (distance to road or building)	A. < 100m	3	5
	B. 100–200m		2
	C. > 200		1
Hydrology (distance to river)	A. ≤ 200m	2	5
	B. 200–400m		4
	C. 400–600m		3
	D. 600–1,000m		2
	E. ≥ 1,000m		1
Lithology/ lithostratigraphic unit	A. Higher Himalayan crystallines	8	9
	B. Schist and gneiss		8
	C. Quartzite		5
	D. Limestone and greywacke		4

where the k-th LUZ is covered by built-up land, or transportation, or cropland. In Shenzhen it is hard to reduce the landslide in some places like those with steep slope, high relative relief, and the places close to the thrust. But by the land use allocation optimization, we can reduce the damage caused by the landslide by putting the built-up land, transportation, and cultivated land far away from the places that are vulnerable to the landslide or maintain a higher landslide risk.

9.2 Constraints at the Overall Land Use Plan Level

Shenzhen is undergoing urbanization along with the increase in built-up areas. From the perspective of the current development trend, sustainable urbanization is required to satisfy economic growth and mitigate urban environmental problems. Four constraints are proposed, with the first three on the areas of built-up land, water body, and cultivated land, respectively, and the last one on the topographical limitations of Shenzhen.

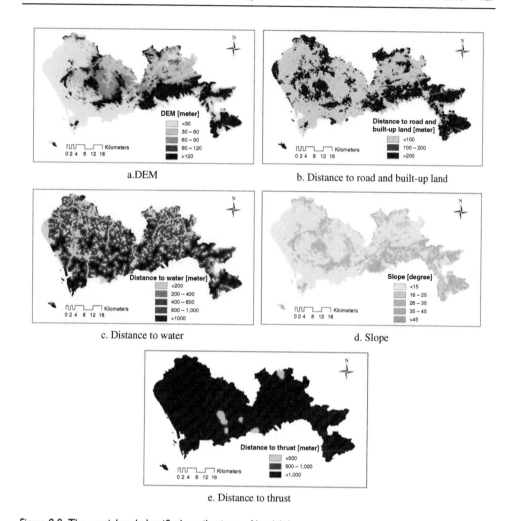

a.DEM

b. Distance to road and built-up land

c. Distance to water

d. Slope

e. Distance to thrust

Figure 9.8 The spatial and classified attributions of landslide

9.2.1 Built-up Land Area

Built-up land area is a significant index for the municipal overall land use plan and the urban master plan. It accommodates the increasing population and provides employment capacity to citizens. However, excessive built-up land may cause environmental problems. Referring to the rules of the municipal overall land use plan and the urban master plan of Shenzhen, as well as the increasing population in the area, we assign constraints on the built-up land area in Shenzhen in 2020.

Population forecasting

The year-end permanent population in Shenzhen from 1979 to 2011 is listed in Table 9.17. According to the definition proposed by National Bureau of Statistics of

Table 9.17 Year-end population of Shenzhen from 1979 to 2011

Year	Year-end permanent population (10,000 persons)	Year	Year-end permanent population (10,000 persons)
1979	31.41	1996	482.89
1980	33.29	1997	527.75
1981	36.69	1998	580.33
1982	44.95	1999	632.56
1983	59.52	2000	701.24
1984	74.13	2001	724.57
1985	88.15	2002	746.62
1986	93.56	2003	778.27
1987	105.44	2004	800.8
1988	120.14	2005	827.75
1989	141.6	2006	871.1
1990	167.78	2007	912.37
1991	226.76	2008	954.28
1992	268.02	2009	995.01
1993	335.97	2010	1,037.2
1994	412.71	2011	1,046.74
1995	449.15		

the People's Republic of China, permanent population includes (1) registered permanent residents living in the city, (2) unregistered permanent residents who have lived in the city for more than half a year, and (3) registered permanent residents of the city who have lived outside the city for less than half a year.

Future built-up areas and other land use objectives are related to population growth. Hence, future population is important for land use plans. Four population forecasting models are proposed based on historical population data to forecast the population in the projected year (Bierens & Hoever, 1985; Sanderson, 1998). The four models are regression models. Time/year functions as the variable in the models, and the four models reflect different population patterns.

The first model is a linear polynomial model (see Table 9.18 & Figure 9.9) and the forecast population in 2020 is 13.6840 million; the second model is a quadratic

Table 9.18 Analysis of the population forecasting model: Linear polynomial

Linear model Poly1:
$$f(x) = p1*x + p2$$

Coefficients (with 95% confidence bounds):
p1 = 36.18 (33.93, 38.43)
p2 = −7.172e+004 (−7.621e + 004, −6.722e + 004)

Goodness-of-fit:
SSE: 1.13e + 005
R-square: 0.9719
Adjusted R-square: 0.971
RMSE: 60.38

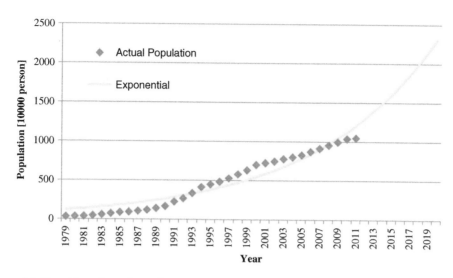

Figure 9.9 Trend line of population forecast by the exponential model

polynomial model (see Table 9.19 & Figure 9.10) and the forecast population in 2020 is 16.0539 million; the third model is an exponential model (see Table 9.20 & Figure 9.11) and the forecasted population in 2020 is 23.1065 million; and the fourth model is a logarithmic model (see Table 9.21 & Figure 9.12) and the forecast population in 2020 is 13.99 million.

Aside from the forecasting models, plans in Shenzhen also have some expectations concerning the future population. The UMP asserted that the population should be no more than 11 million in 2020 (Shenzhen Municipal Government, 2010). In addition, the 12th Five-Year Plan of National Economic and Social Development of Shenzhen (Shenzhen Municipal People's Government, 2011) ruled that population should be no more than 11 million in 2015.

Table 9.19 Analysis of the population forecasting model:Quadratic polynomial

Linear model Poly2: $f(x) = p1*x^2 + p2*x + p3$

Coefficients (with 95% confidence bounds):
  $p1 = 0.4435$ (0.2308, 0.6563)
  $p2 = -1733$ (-2582, -884.5)
  $p3 = 1.693e + 006$ (8.467e + 005, 2.54e + 006)

Goodness-of-fit:
  SSE: 7.046e + 004
  R-square: 0.9825
  Adjusted R-square: 0.9813
  RMSE: 48.46

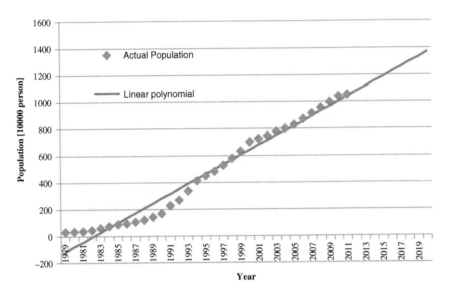

Figure 9.10 Trend line of population forecast by the linear polynomial model

Table 9.20 Analysis of the population forecasting model: Exponential

General model Exp1:   f(x) = a*exp(b*x)

Coefficients (with 95% confidence bounds):
  a = 1.553e − 060 (−2.831e − 059, 3.142e − 059)
  b = 0.07201 (0.06242, 0.0816)

Goodness-of-fit:
  SSE: 2.867e+005
  R-square: 0.9288
  Adjusted R-square: 0.9266
  RMSE: 96.17

All the expected populations by forecasting models and plans in Shenzhen are summarized in Table 9.22.

Built-up per capita

Table 9.23 lists the per capital built-up land in Shenzhen from 1996 to 2008. The total area of built-up land obviously increased from 434.85km^2 to 754.43km^2. The population increased from 4.8289 to 9.5428 million during the same period. From 1996 to 2000, the population growth rate was larger than the rate of built-up land increase rate. Therefore, the per capita built-up land decreased. After 2000, the increasing rates of population growth and built-up land area became parallel. Then, per capita built-up land became stable at approximately 76m^2/person to 80m^2/person. Based on the

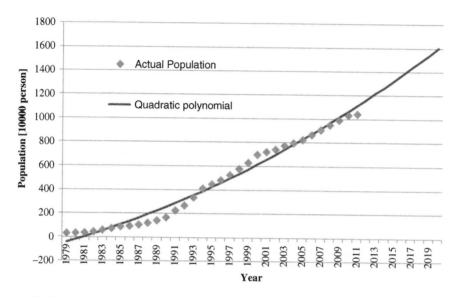

Figure 9.11 Trend line of population forecast by the quadratic polynomial model

Table 9.21 Analysis of the population forecasting model: Logarithmic.

Linear model Poly I: $f(x) = p1*\ln(x) + p2$
Coefficients (with 95% confidence bounds): $p1 = 7.216e + 004$ (6.764e + 004, 7.669e + 004) $p2 = -5.478e + 005$ (−5.822e + 005, −5.135e + 005)
Goodness-of-fit: SSE: 1.148e + 005 R-square: 0.9715 Adjusted R-square: 0.9706 RMSE: 60.85

temporal variation of per capita built-up land from 1996 to 2008, the value will be approximately 76m²/person to 80m²/person in future, which satisfies the rules proposed by the Ministry of Housing and Urban–Rural Construction of the People's Republic of China (2011) (see Table 9.24).

According the rules proposed by the Ministry of Housing and Urban–Rural Construction of the People's Republic of China (2011) the per capita built-up land in Shenzhen was 79.06m²/person in 2008. In the future plan, per capita built-up land is allowed to range from 75.0m²/person to 105.0m²/person. However, according to the municipal overall land use plan of Shenzhen, the per capita built-up land should be smaller than 81m². Table 9.25 lists the expected per capita built-up land in Shenzhen in 2020 by using different methods or policies.

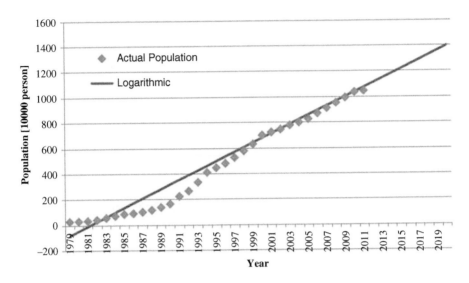

Figure 9.12 Trend line of population forecast by the logarithmic model

Table 9.22 Summary of all the expected populations by forecasting models and plans in Shenzhen

Model or method	Forecast population [10,000 persons]	Year
Linear polynomial model	1,368.40	2020
Quadratic polynomial	1,605.39	2020
Exponential model	2,310.65	2020
Logarithmic model	1,399.14	2020
Urban master plan of Shenzhen (2010–2020)*	< 1,100	2020
The 12th five year plan of national economic and social development of Shenzhen	< 1,100	2015

* Source: Shenzhen Municipal Government (2010)

Table 9.23 Per capita built-up land of Shenzhen from 1996 to 2008

Year	1996	2000	2002	2004	2006	2008
Built-up land (km^2)	434.85	489.03	571.71	642.95	676.92	754.43
Population (10,000 persons)	482.89	701.24	746.62	800.8	871.1	954.28
Built-up area per capita (m^2/person)	90.05	69.74	76.57	80.29	77.71	79.06

Expected built-up land in the projected year

Both the municipal overall land use plan and urban master plan have some require-ments on the built-up land in 2020. The MOLUP and the UMP deem that the total area of built-up land should be 990 km^2 and 890km^2 in 2020, respectively. Aside from the expectations of the MOLUP and the UMP, the total area of built-up land in the

Table 9.24 Allowed built-up land per capita in plan in m2/person*

Current built-up land per capita	The allowed the built-up land per capital
≤ 65	65.0 ~ 85.0
65.1 ~ 75.0	65.0 ~ 95.0
75.1 ~ 85.0	75.0 ~ 105.0
85.1 ~ 95.0	80.0 ~ 110.0
95.1 ~ 105.0	90.0 ~ 110.0
105 ~ 115.0	95.0 ~ 115.0
> 115.0	≤ 115.0

Source * Ministry of Housing and Urban–Rural Construction of the People's Republic of China (2011)

Table 9.25 Expected per capita built-up land of Shenzhen in 2020 by different methods

Methods or policies	Expected per capita built-up land [m²/person]
Historical tendency	76–80
Rules in Code for classification of urban land use and planning standards of development land*	75.0–105.0
Rules in the overall land use plan of Shenzhen (2006–2020)**	78
Rules in the urban master plan of Shenzhen (2010–2020)***	< 81

Source * Ministry of Housing and Urban–Rural Construction of the People's Republic of China (2011) ** Urban Planning Land and Resources Commission of Shenzhen Municipality (2011): *** Shenzhen Municipal Government (2010)

projected year can be determined according to the forecast population and per capita built-up land.

The forecast population is 13.99 million, and the per capita built-up land is determined as 78m²/person, which satisfies all the policies and the historical tendency. Therefore, the expected area of built-up land is 13.99 million × 78 m²/person = 1,091.33km². By contrast, the population expected by the plans is smaller than 11 million. In this case, the expected area of built-up land is 11 million × 78m²/person = 858km². All the expected areas of built-up land are listed in Table 9.26. Accordingly, the area of built-up land should range from 858km² to 1,091.33km² in the projected year. It is set as one constraint on the area of built-up land. By the MOO, a set of Pareto-optimal solutions is generated to maintain different built-up land areas. Planners or decision-makers can select from these solutions.

9.2.2 Water Body

Water bodies should not be developed because of the requirements of ecological protection. Therefore, water bodies should be 141.34km² in 2020, which will be the same as that in 2008.

Table 9.26 Expected area of built-up land in 2020 proposed by different models

Approach or model	Expected area of built-up land in 2020 [km^2]
The municipal overall land use plan of Shenzhen	990
The urban master plan of Shenzhen	890
By the forecasted population and per capital built-up land	1,091.33
By the expected population of plans and the per capital built-up land	858

9.2.3 Cultivated Land

For the foreseeable future, China needs to regard the protection of farmlands as the key point because of the significant demand of food supply from the country's large population. The Regulations on Basic Farmland Protection (The State Council, 1999) listed the detailed contents of farmland protection in China. This law guarantees the protection of farmlands in China.

In particular, the required basic farmland in Shenzhen assigned by planning is 64.64km^2. That is, the farmland in Shenzhen should be no less than 64.64km^2. In 2004, the Department of Land and Resources of Guangdong Province formally approved the placement policy of basic farmlands in Shenzhen (Urban Planning Land and Resources Commission of Shenzhen Municipality, 2011). Through this policy, Shenzhen exhibits the right to ask another city to provide a 40km^2 basic farmland for Shenzhen and then give economic compensation to this city. Therefore, the required basic farmland in Shenzhen will be reduced to 24.65km^2 in 2020. The value of 24.65km^2 is set as the constraint on the area of cultivated land in Shenzhen.

9.2.4 Elevation

Shenzhen is ecological and is difficult to be developed because of its mountainous topography. In particular, the municipal overall land use plan of Shenzhen (2006–2020) (Urban Planning Land and Resources Commission of Shenzhen Municipality, 2011) ruled that lands with elevations above 80m are prohibited from being developed.

9.2.5 Reserve Protection

There is a national nature reserve in Shenzhen, namely, the Futian National Nature Reserve. According to local policy, the reserve will not be developed in the future, and therefore it functions as a constraint.

Chapter 10

Optimization at the Urban Master Plan Level

In this chapter, the detailed definitions and measurements of the objectives at the urban master plan level are proposed at the outset. They concentrate on developments in the urban areas rather than those in the municipal areas. According to basic principles, the proposed objectives include the maximization of housing capacity, employment capacity, mixed land uses, green space, accessibility, compatibility, and spatial equity, and the minimization of the changing cost and the pollution from industrial land on residential land. Secondly, the constraints relate mainly to the limitations of various land use areas, including the proportion of residential land, public service land, industrial land, and green land.

10.1 Objectives at the Urban Master Plan Level

Futian, which is one of the SEZs, is selected as an example of urban land use optimization. The urban land use pattern of Futian in 2008 is mapped in Figure 10.1, wherein the northern part of the district is covered mainly by non-urban land uses and the southern part is covered mainly by urban land uses.

The possible urban land usages in Shenzhen and the corresponding abbreviations and index IDs are listed in Table 10.1. In an urban system, housing and employment capacities should be sufficient to satisfy the needs of the increasing population. Secondly, the environment in the urban areas should be sufficiently comfortable to maintain the quality of life in a city. The entire urban system should also be accessible and equitable to its citizens. The objectives of urban land use are defined and measured as follows.

10.1.1 To Maximize Housing Capacity

According to possible urban land usages, residential land can be cataloged into first, second, or third class. According to literature review as well as the historical residential land and population data in Shenzhen, the total housing capacity of candidate plan i can be computed by multiplying the residential land area with the unit housing capacity in people per hectare for individual land use (CNumRP) (Balling et al., 2004; Chandramouli et al., 2009). The first objective can be computed by Equation 10.1:

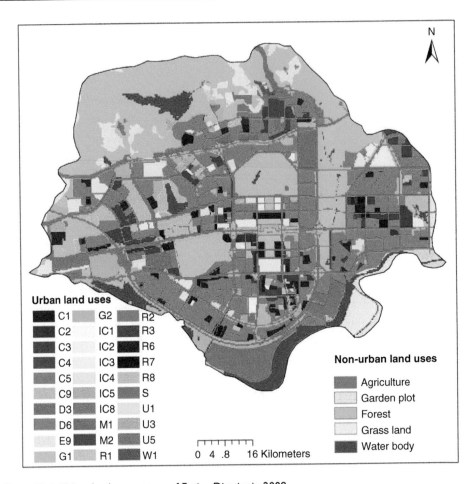

Figure 10.1 Urban land use pattern of Futian District in 2008.

$$max\ Z^{u}_{NumRP} = \sum_{k=1}^{Num_{ulu}} Area_{k} * CNumRP_{k} \tag{10.1}$$

where Z^{u}_{NumRP} is the housing capacity, and $Area_{k}$ is the area of the k-th land usage, $CNumRP_{k}$ is unit housing capacity in people per hectare for k-th land usage.

The unit housing capacity, which exerts a considerable effect on the final housing capacity for the optimal plan in Futian, is important in Equation 10.1. The population ratio of Shenzhen in 2007, which is 9123700, available from the statistic yearbook of Shenzhen (2008) (Shenzhen Statistics Bureau, 2008), to the area of residential land in 2007, which is 19,690 hectares, retrieved from UMP, is 463.37 persons per hectare. Therefore, one hectare of residential land can on average accommodate 463.37 people in Shenzhen.

Retrieving detailed statistical data to determine the unit housing capacities of first-, second-, and third-class residential lands, we assume that the unit housing capacities of

Table 10.1 Possible urban land uses

ID	Urban land uses	A.b.	Urban land uses
1	Commerce	C1	Commercial land
2		C2	Banking land
3		C3	Service industry land
4		C4	Hotel industry land
5		C5	Commercial office land
6		C9	Tourism and recreation land
7		D3	Exempt zone
8		D6	Nature reserve land
9		E1	Water body
10		E5	Forest
11		E9	Development of backup land
12	Green space	G1	Green space and square
13		G2	Production buffer green area
14	Public management and public service land	IC1	Administrative land
15		IC2	Culture and entertainment land
16		IC3	Sports facility land use
17		IC4	Medical hygienic land
18		IC5	Education and scientific research land
19		IC8	Port land
20	Industrial land	M1	First-class industrial land
21		M2	Second-class industrial land
22	Residential land	R1	First-class residential land
23		R2	Second-class residential land
24		R3	Third-class residential land
25		R6	Residential facilities land
26		R7	Residential parking land
27		R8	Residential green space
28		W1	Logistics and warehouse land
29		S	Transportation and utilities land
30	Municipal utilities land	U1	Supplying utilities land
31		U3	Post and telecommunications utilities land
32		U4	Environmental sanitation utilities land
33		U5	Construction and repair utilities land

these lands are 520, 463.37, and 100 persons per hectare, respectively. This assumption is based on the characteristics of these residential lands, as ruled by the Ministry of Housing and Urban–Rural Construction of the People's Republic of China (2011). Other urban land uses in Shenzhen do not generate any housing capacity. The unit housing capacity of each urban land usage is listed in Table 10.2.

10.1.2 To Maximize Employment Capacity

The second objective is proposed to maximize the employment capacity for the residents. Different land uses vary in unit employment capacities in people per hectare.

Table 10.2 Unit housing capacities in person per hectare of different land uses

ID	A.b.	CNumRP	ID	A.b.	CNumRP
		persons per hectare			persons per hectare
1	C1	0	18	IC5	0
2	C2	0	19	IC8	0
3	C3	0	20	M1	0
4	C4	0	21	M2	0
5	C5	0	22	R1	100
6	C9	0	23	R2	463.37
7	D3	0	24	R3	520
8	D6	0	25	R6	0
9	E1	0	26	R7	0
10	E5	0	27	R8	0
11	E9	0	28	W	0
12	G1	0	29	U1	0
13	G2	0	30	S	0
14	IC1	0	31	U3	0
15	IC2	0	32	U4	0
16	IC3	0	33	U5	0
17	IC4	0			

Using existing studies as well as historical employment capacity and the land use data (Balling *et al.*, 2004; Chandramouli *et al.*, 2009), the CNumEploy for different land usage are assigned. Then, the objective of maximizing the employment capacity can be calculated by Equation 10.2:

$$max\ Z^u_{NumEploy} = \sum_{k=1}^{Num_{ulu}} Area_k * CNumEploy_k \qquad (10.2)$$

where $Z^u_{NumEploy}$ is the employment capacity, $Area_k$ is the area of k-th land usage, $CNumEploy_k$ is the unit employment capacity in people per hectare for k-th land usage, Num_{ulu} is the number of possible urban land use type.

$CNumEploy_k$ is the key point in determining the employment capacity of Futian. Assigning the unit employment capacity on each urban land usage by referring to the coefficients of other cities is difficult because of their different economies, human resources, and local policies. Therefore, the area of each urban land usage in 2007 and the employed people in Shenzhen in 2007 grouped by the standard industrial classification sector (SIC) are used herein to calculate the CNumEploy.

The number of employed individuals in Shenzhen grouped by sector in 2007 is available from the 2008 Shenzhen yearbook (Shenzhen Statistics Bureau, 2008). The corresponding land use of each sector can be determined by checking the contents of the Standard Industrial Classification Codes (2002) and the Code for Classification of Urban Land Use and Planning Standards of Development Land which define the detailed uses of each urban land usage. The former explains the concrete contents of each sector, whereas the latter defines the detailed uses of each urban land usage. Even if these two classification standards are different, urban land usage is determined by the

Table 10.3 Number of employed of Shenzhen grouped by sector in 2007

Grouped by sector	Corresponding urban land use type	Total employed (year-end) *
Farming, forestry, animal husbandry and fishery	Non-urban land usages	6,972
Mining and quarrying	Industrial land	1,339
Manufacturing	Industrial land	3,356,170
Electricity, gas and water production and supply	Municipal utilities land	19,613
Construction	All urban land usages	169,558
Transportation, storage and post services	Transportation and utilities land + logistics and warehouse land	162,889
Information transfer, computer and software services	Commercial and service facility land	134,078
Wholesale and retail sales	Commercial and service facility land	1,070,436
Accommodation and catering trade	Commercial and service facility land	325,308
Banking	Commercial and service facility land	69,697
Real estate trade	Commercial and service facility land	317,949
Leasing industry and commercial services	Commercial and service facility land	269,789
Scientific research, technical services and geological prospecting	Commercial and service facility land	86,561
Water conservancy, environment management and public amenities	Municipal utilities land	43,364
Resident services and other services	Commercial and service facility land	227,806
Education	Education and scientific research land**	104,011
Health care, social insurance and welfare	Public management and public service land	46,563
Culture, sports and entertainment	Culture and entertainment land** + sports facility land Use**	41,887
Public services and social organizations	Public management and public service land	101,847
International organizations	Public management and public service land	0

Source: * Shenzhen Statistics Bureau (2008); Shenzhen Statistics Bureau (2008); ** belonging to "Public management and public service land"

activities on or the uses of land. Moreover, the Standard Industrial Classification Codes classify the activities from the aspect of the economy. Therefore, the corresponding urban land use type for the standard industrial classification sectors can be identified (see Table 10.3).

The structure of urban land usages in 2007 is available in the UMP (see Table 10.4). Unit employment capacity is determined through the ratio of the number of employed people in a given urban land usage to the area of the given urban land usage.

People employed in the construction sector work in the entire urban area but not on a specific urban land usage because construction occurs everywhere. The unit employment capacity of construction is calculated from the ratio of the number of employed people in the construction sector (169,558 persons) to the total urban area in 2007 (75,030 hectares). Therefore, the unit employment capacity for the construction sector is determined as 2.26 person per hectare in urban areas, which is added to the unit

Table 10.4 Employment capacity for each urban land usage of Shenzhen in 2007

A.b.	Urban land use	Area* (ha)	Total employed (excluding construction)	Unit employment capacity excluding construction	Unit employment capacity of construction	Unit employment capacity
				[persons per hectare]		
R	Residential land	19,690	0	0.00	2.26	2.26
C	Commercial and service facility land	3,590	250,1624	696.83	2.26	699.09
IC	Public management and public service land	3,960	294,308.00	74.32	2.26	76.58
IC	Education and scientific research land	2,054	104,011	50.64	2.26	52.90
in which	Medical hygienic land	416	46,563	119.93	2.26	114.19
	Culture and entertainment land	314	41,887	70.28	2.26	72.54
	Sports facility land use	282		70.28	72.54	72.54
M	Industrial land	26900	3,357,509	124.81	2.26	72.54
W	Logistics and warehouse land	1210	162,889	12.08	2.26	127.07
S	Transport and utilities land	12270		12.08	2.26	14.34
U	Municipal utilities land	2060	62,977	30.57	2.26	14.34
G	Green space	4860	0	0.00	2.26	32.83
D	Special land use	490	0	0.00	2.26	2.26
Sum	Urban area	75030	0	–	–	87.28

* source: UMP

Table 10.5 Unit employment capacities in people per hectare of different urban land uses

ID	A.b.	CNumRP	ID	A.b.	CNumRP
		persons per hectare			persons per hectare
1	C1	699.09	18	IC5	52.90
2	C2	699.09	19	IC8	76.58
3	C3	699.09	20	M1	127.07
4	C4	699.09	21	M2	127.07
5	C5	699.09	22	R1	2.26
6	C9	699.09	23	R2	2.26
7	D3	87.28	24	R3	2.26
8	D6	0	25	R6	2.26
9	E1	0	26	R7	2.26
10	E5	0	27	R8	2.26
11	E9	0	28	W1	14.34
12	G1	0	29	U1	32.83
13	G2	0	30	S	14.34
14	IC1	76.58	31	U3	32.83
15	IC2	72.54	32	U4	32.83
16	IC3	72.54	33	U5	32.83
17	IC4	2.26			

employment capacities for all land usages, excluding construction, to obtain the final unit employment capacities (see Table 10.4).

The unit employment capacities of different urban land usages in Shenzhen are determined from the results listed in Table 10.4. It assumes that Futian maintains the same unit employment capacities as Shenzhen. Land use type D3 (Exempt zone) in Futian functions as the special zone in the district, wherein commercial, industrial, and all other kinds of urban land usages are found. Considering that determining the specific spatial allocation of urban land usage in this zone is difficult, it assumes that the exempt zone maintains the same unit employment capacity as that of the entire urban area in Futian, which is 87.28 persons per hectare (see Table 10.5).

10.1.3 To Minimize the Changing Cost

The redevelopment of developed urban lands may be costly. In this case, changing cost is defined as the population that is affected by redevelopment in accordance with the study by Balling *et al.* (2004). Then, minimizing the changing cost is defined as one of the objectives, as follows (see Equation 10.3):

$$max\ Z^{u}_{ChangingCost} = \sum_{k=1}^{n} Area_k * (CNumEploy_k + CNumRP_k) \qquad (10.3)$$

where n is the number of urban land use type, and $Area_k$ is the k-th urban land usage that has been changed.

10.1.4 To Minimize the Pollution from Industrial Land on Residential Land

Wind significantly influences urban climate (Grimmond & Souch, 1994). The diffusion of industrial, commercial, institutional, and residential activities, as well as the movement of vehicle emissions in a city, are related to wind direction and frequency. Understanding the relation between urban morphology and favorable wind conditions in a city is important in order to avoid serious urban air pollution episodes and energy consumption, thereby improving the quality of life in urban environments. Manolopoulos *et al.* (2007) studied the sources of speciated atmospheric mercury in a residential neighborhood affected by industrial sources and found that wind direction affects pollution diffusion.

In fact, the dominant wind direction has long been used to determine the location of residential lands in the UMP, wherein residential land should be located in the upwind direction (Schmauss, 1914). This classic method is concerned with the dominant wind direction and it attempts to allocate residential land in the upwind direction to avoid diffused pollution from industries. In reality, however, wind direction constantly varies.

Therefore, the wind rose chart is used in this publication to determine the corresponding wind frequency in all directions. Wind frequency is initially normalized, and then wind frequencies in all directions range from 0 to 1.

The air pollution index for industrial land can be determined by calculating the angle from industrial land to residential land. Angle, θ_i, from industrial land, I, to residential land, R_i, can be defined as the angle of the gravitational centers of industrial land (x_I, y_I) and residential land (x_{Ri}, y_{Ri}) (see Figure 10.2 & Equation 10.4). Wind frequency can be determined from the angle and the corresponding wind rose chart. Figure 10.3 shows that when residential land is located around 300 degrees, the wind frequency is 0.245.

$$\theta_i = \operatorname{atan}\left(\frac{x_{Ri} - x_I}{y_{Ri} - y_I}\right) \tag{10.4}$$

This publication focuses only on the effect of industrial pollution on residential land. It sums the wind frequencies for each residential land from the first to the last. The effect

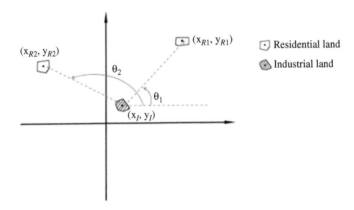

Figure 10.2 The diagram of the angle of industrial lands to residential lands

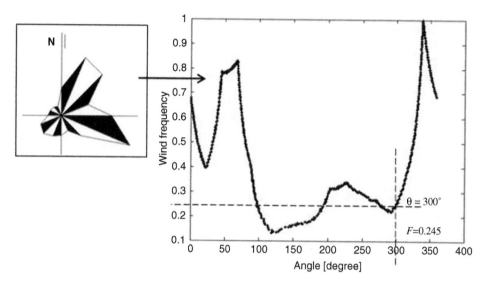

Figure 10.3 The wind frequency in different directions

of air pollution from a particular industrial land to all residential lands can be defined by summing the following values:

$$\min Z^u_{\text{AirPollution}} = \sum_{j=1}^{\text{Num\_M}} \sum_{k=1}^{\text{Num\_R}} \text{Area}_k \text{windf}_{jk} \qquad (10.5)$$

where $Z^u_{\text{AirPollution}}$ is the air pollution index, Area_k is the area of the k-th residential land use zone, windf_{jk} is the wind frequency of the j-th industrial land use zone to the k-th residential land use zone, and Num_M and Num_R are the numbers of industrial and residential land use zones, respectively.

Figure 10.4 shows two examples wherein two sets of industrial land are allocated, along with the wind frequency in which the industrial land is considered to be a pollution source. All ULUZs in Futian are colored. The color indicates the degree of

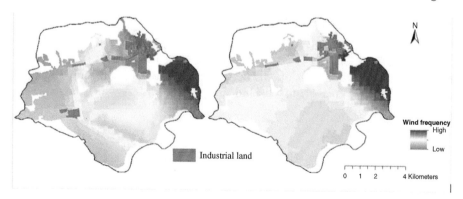

Figure 10.4 Two examples of wind frequencies from the source of industrial land

pollution effect from an industrial land. The map shows that all lands are affected by pollution because wind changes direction throughout the year. However, the pollution degree varies in different directions because of the location of industrial lands and the frequency of wind vary. The optimization process aims to properly allocate industrial land and residential land to ensure that pollution in the residential land from the industrial land is relatively light.

10.1.5 To Maximize Mixed Land Uses

Mixed land use is the backbone of contemporary planning policies, such as smart growth and new urbanism (Koster & Rouwendal, 2012). Despite the wide variety of definitions, the concept generally focuses on mixed neighborhoods with a relatively high density in terms of land use. Numerous methods can be used to measure whether land use is mixed. Referring to Yamada *et al.*'s 2012 study, Frank's entropy score, Shannon's index, and Simpson's index can be used to measure the mixed land use as Equation 10.6 and 10.7 show:

$$M_{mix\_landuse} = -\sum_{i=1}^{6} \left(b_i \Big/ a \right) * \ln \left(b_i \Big/ a \right) / ln(N) \tag{10.6}$$

$$M_{mix\_landuse} = -\sum_{i=1}^{6} (p_i) * \ln(p_i) / ln(N) \tag{10.7}$$

where $M_{mix_{landuse}}$ is the degree of mixed land use, i = 1, 2, 3... denote several certain kinds of urban land use types; a is total area of land for a neighborhood; N is the number of certain land usages with area > 0; b_i is the area of the i-th land usage; and $p_i = b_i \big/ a$ is percentage of area of the i-th land usage. Six types of urban land usages are generally considered: single-family residential; multifamily residential; retail; office; education; and entertainment (Yamada *et al.*, 2012). The urban land use classification system adopted in this publication is different from those employed in previous studies. Thus, R, G, IC, M, U, and S are selected to measure mixed land use in Shenzhen. Neighbor is defined by the spatial weights matrix in this study. Therefore, the mixed land use for a candidate plan can be defined as follows by summing all mixed land use indices of each ULUZ:

$$max \; Z^u_{mixed\_landuse} = \sum_{k=1}^{num\_NZ} Area_k M_{mix\_landuse,k} \tag{10.8}$$

where num_{NZ} is the number of ULUZ, $Area_k$ is the area of k-th ULUZ, and $M_{mix\_landuse,k}$ is the degree of mixed land use of k-th ULUZ which is achieved by Equation 10.7.

10.1.6 To Maximize Green Space

The objective of maximizing green space requires a future plan that provides sufficient green space for all citizens. The green space of a particular plan is achieved by summing the areas of green space ULUZs, as follows:

$$max\ Z^u_{GS} = \sum Area_G \tag{10.9}$$

where Z^u_G is the total area of green space and square, and $Area_G$ is the area of a certain LUZ which is covered as green space and square.

10.1.7 To Maximize Accessibility

Accessibility is defined and implemented in several ways, and therefore it can be determined through a variety of methods. For example, accessibility can be defined as the potential for interaction opportunities (Hansen, 1959), or the ease with which any land use activity can be reached from a location by using a particular transport system (Dalvi & Martin, 1976). Consequently, accessibility at the MOLUP level is defined as the accessibility of a road network, which measures the convenience of reaching the nearest road network. By contrast, other scholars regard accessibility as the freedom of individuals to participate in different activities (Burns, 1980), or the benefit provided by a transportation/land use system (Akiva & Lerman, 1985). In summary, there are various means by which accessibility is measured.

Accessibility should also be defined at the UMP level. This publication focuses on the spatial layout and structure of land use. Thus, measuring the accessibility of benefits provided by land usages is reasonable. In addition, location-based accessibility is frequently used in land use planning, as follows (Geurs & Van Wee, 2004):

$$max\ Z^u_{accessbility} = \sum_{i=1}^{LUZ_i \in U_{urban}} \sum_{j=1}^{Dis_{ij} < D} \frac{1}{Area_i} * \frac{1}{Dis_{ij}} * Area_j * Acc_j \tag{10.10}$$

where $Area_i$ and $Area_j$ are the area of i-th and j-th ULUZs, Dis_{ij} is the distance between i-th ULUZ and j-th ULUZ, and Acc_j denotes the benefit provided by j-th ULUZ named as the accessibility benefit index. For example, the benefit could be job opportunity, education, or open space.

Different urban land uses require various accessibility benefits. Thus, C, U, IC, and G are set to provide benefits to R; R, U, IC, and G to C; R, G, IC, U, and C to M; and R, G, IC, U, and C to W. M causes various kinds of air, water, and noise pollution; thus, M is not considered to be a beneficial land use. Taking R as an example, people living in R need job opportunities (provided by C, U, and IC, where C is the major one), shopping (provided by C), entertainment (provided by G and IC), and infrastructure. Therefore, C, U, IC, and G are defined as the beneficial land for R. In particular, the amount of benefits of a given beneficial land is measured by the area of a given beneficial land.

10.1.8 To Maximize Compatibility

Similar to LUZs at the level of municipal overall land use plan, the ULUZ has its own preference to choose a land use to act as its neighborhood. In this study, compatibility merely indicates the relationship between the ULUZs and not the relationship between the LUZs and ULUZs. In particular, this study defines the compatibility index between every two urban land use types, which ranges from 0 to 1, with a high value that indicates high compatibility between the two land usages. The compatibility of a candidate plan can be indicated by adding all the compatibility indices between each plot and its neighborhood. In addition, a given ULUZ with a large area exhibits high conflict with its neighbor. Therefore, the sum of the compatibility index and the areas of a given ULUZ and its neighbor indicates compatibility. When the indicator is large, the scenario is compatible. In setting the compatibility index, the opinions of experts, decision-makers, and stakeholders in land use allocation must be considered. Then, the compatibility of the proposed plan can be formulated by Equation 10.11:

$$\max Z_{com} = \sum_{i=1}^{ULUZ_i \in U_{urban}} \sum_{j=1}^{ULUZ_j \in NB_i} ComInd_{k,m} \times Area_i \times Area_j \times w_{i,j} \qquad (10.11)$$

Where Z_{Com} is the compatibility of a certain plan; NB_i is the neighborhood of the i-th ULUZ; $Area_i$ and $Area_j$ are the area of i-th ULUZ and j-th ULUZ; $ComInd_{k,m}$ is the compatibility index between the k-th land use type and the m-th land use type; and $w_{i,j}$ is the spatial weight between i-th ULUZ and j-th ULUZ.

Futian has 34 urban land usages. Considerable efforts must be made to determine the compatibility index between each two urban land usages. Similar to the process of extracting the compatibility index at the overall land use plan level, the AHP is used to extract opinions from the planners and decision-makers. The AHP is a multiple criteria decision-making tool and it has been used in almost all of the the applications related with decision-making (Vaidya & Kumar, 2006). Since the AHP helps to incorporate a group consensus, it can be used in the process of extracting opinions of the interviewers in the process of determining the compatibility index for urban land usages. Firstly, the concept and meaning of compatibility should be introduced to the interviewers who are planners, decision-makers, and experts. Also some compatibility indexes that have been proposed in existing studies (Ligmann-Zielinska et al., 2005; Ligmann-Zielinska et al., 2008; Cao et al., 2012) have been shown to the interviewers. Then, it needs to establish a two-level analytic hierarchy model. At first level, it includes five elements: C (C1, C2, C3, C4, C5, C9), D (D3), IC (IC1, IC2, IC3, IC4, IC5, IC8), M (M1, M2), R (R1, R2, R3, R6, R7, R8), W (W1), U (U1, U3, U4, U5). The interviewers are asked to evaluate the compatibility index between these five elements. And, then, at the second level, the interviewers are asked to evaluate the compatibility index for the land uses within every two C, IC, M, R, or U. Finally, the results at the two levels are combined.

10.1.9 To Maximize Spatial Equity

Due to the preferential policies for the SEZs, there is a big gap between the SEZs and the other districts in Shenzhen. A good land use pattern or the urban master plan provides equitable access to social and economic resources. Therefore, the equity is herein defined as one objective. Equity is the quality of being equal or fair (Alonso, 1968). Since the urban land use optimization will be carried out just for one district, that of Futian , the equity of Futian rather than the equity of the whole city is concerned.

Many benefits variables can be considered when measuring equity. For example, Landry and Chakraborty (2009) evaluated the spatial equity considering the street trees. Talen and Anselin (1998) assessed the spatial equity in accessibility to public playgrounds. In this publication, four benefits variables, namely, housing capacity, employment capacity, green space, and accessibility, are here considered to reflect the spatial equity in Futian. There are many indexes which can be used to measure spatial equity, such as the Schutz index, the Lorenz curve, and the Gini coefficient (Maruyama & Sumalee, 2007). Because the Schutz index is the best simple index for measuring spatial equity (Gaile, 1984) and it has been widely used (Truelove, 1993), it is selected in this study as the index for measuring the spatial equity in Futian, and it is given by:

$$z = \sum \left| \frac{100x_i}{\sum x_i} - \frac{100}{n} \right| \tag{10.12}$$

where x_i is the measure of the benefits variable for the i-th ULUZ, n is the number of ULUZs, and z is Schutz index. The values calculated for z must be interpreted on an undimensional scale of 0 to 200. A lower value indicates an equitable distribution. Then, the objective of maximizing the spatial equity for Futiancan be written as follows:

$$\min Z_{\text{Equity}}^u = \sum_{k=1} z_k \tag{10.13}$$

where z_k is Schutz index for k-th benefits variable, including housing capacity, employment capacity, green space, and accessibility.

10.2 Constraints at the Urban Master Plan Level

In addition to the objectives at the urban master land level, some constraints on urban land use structure are proposed at the urban master plan level.

According to the UMP, the ratios of these four kinds of land usages, namely, (1) residential, (2) administration and public service, (3) industrial, and (4) green space and square, should satisfy the following requirements (see Equation 10.14 to Equation 10.17). It assumes that the land use structure in Futian should be ruled by the UMP.

$$25\% \le \frac{\text{Area}_R}{\text{Area}_{\text{urban}}} \le 40\% \tag{10.14}$$

$$5\% \le \frac{\text{Area}_{\text{APS}}}{\text{Area}_{\text{urban}}} \le 8\% \tag{10.15}$$

$$15\% \le \frac{\text{Area}_{\text{HI}} + \text{Area}_{\text{LI}}}{\text{Area}_{\text{urban}}} \le 30\% \tag{10.16}$$

$$10\% \le \frac{\text{Area}_{\text{GS}}}{\text{Area}_{\text{urban}}} \le 15\% \tag{10.17}$$

Chapter 11

Model Implementation and Evaluation

This chapter introduces a detailed application example on multi-objective land use optimization at two levels. The process of a genetic algorithm (GA) is firstly presented here. It is the tool used to search for the optimal solutions in the MOO approach. After that, the detailed implementation is carried out, including data organization and the optimization of the model implementation. The third part analyzes the optimization results at two levels, and validates the effectiveness of the MOO and its practicability in space.

11.1 Multi-objective Optimization by Genetic Algorithm

A general multi-objective optimization problem comprises a number of objectives associated with a number of inequality and equality constraints (Srinivas & Deb, 1994). A common difficulty in multi-objective problems is the existence of objective conflict. This means that none of the feasible solutions allows simultaneous optimal solutions for all of the objectives. In other words, individual optimal solutions of each objective are usually different.

A GA-based multi-objective optimization technique is selected to search for optimal solutions. The vector map is used in the process of optimization where land use zones act as genes in the GA. Each gene will be set as an integer, ranging from 1 to the number of possible land uses.

The Pareto-optimal solution in the MOO problem means that an improvement in one objective must come at the expense of at least one of the other objectives (Steuer, 1989; Batty, 1998; Miettinen, 1999; Gabriel et al., 2006). These generative techniques allow a multiple scenario analysis, where the outcomes achieved are non-inferior (Pareto-optimal) to the objectives contained in the model. In essence, for a given set of model objectives, it cannot improve the outcome of all objectives without compromising the other objectives. Therefore, the analysis focuses on those options that are not dominated by any other alternatives (Ligmann-Zielinska et al., 2008).

The framework of the GA is illustrated in Figure 11.1. Usually, the GA started with generating a certain number of plans randomly in the first generation, and the number of plans is the population size of the GA. Then, the plans satisfying constraints would be ordered by their fitness. The plans in the second iteration are obtained via the processes of selection, crossover, mutation, and elitism. The same processes are utilized to generate the following generation until the improvement of average fitness of each generation is smaller than a certain threshold or the iteration has achieved a relatively

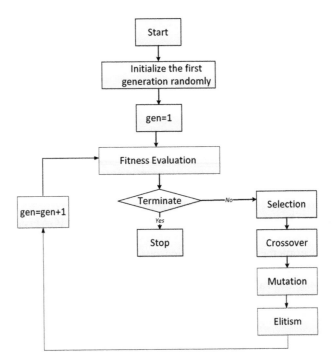

Figure 11.1 Framework of a GA

large number. The detailed processes of selection, crossover, mutation, and elitism are represented as follows.

Selection

Two plans with high fitness are randomly selected from generated feasible plans, named as father and mother.

Crossover

Genes in the parents are exchanged via a crossover to generate children in the next generation.

Mutation

Then, in order to avoid local optimum, the mutation is conducted. The mutation probability is performed on all of the children plans. The mutation probability maintains a very small value ranging from 0 to 1. It means, for each gene, there is a probability as large as the mutation probability to change to other land use value randomly.

Elitism

In order to maintain the good quality from the last generation, about 10% of plans with the highest fitness in the last generation would be maintained in the next generation.

All of these processes are conducted until plans of population size are generated.

11.2 Optimization Implementation

11.2.1 Optimization Unit

This research uses vector data to perform the optimization. A vector data model is a representation of the geographic world by the use of points, lines, and polygons. By contrast, a raster data model is a representation of the geographic world by dividing the surface into a regular grid of cells. Although raster data maintain simple data structure and spatial operations are easily and efficiently implemented, the sheer volume of data to be stored and handled can be very high, and this leads to inefficient optimization. Therefore, vector data are used in this research. Specifically, vector data of land use zones, LUZs, and urban land use zones, ULUZs, divided by road networks are described by using contour lines. The divided LUZs are shown as polygons in the GIS system and act as genes in the GA, during the optimization process. Some geospatial data are in raster format; hence, raster data are converted into vector data. The mean value of a set of grids within the boundary of a certain LUZ is extracted as the attribution of this LUZ. The grid that is simultaneously covered by two LUZs will be divided into these two LUZs to accord with the area in each LUZ.

In previous studies, only the road network is used to divide the case study area into traffic analysis zones (Balling *et al.*, 2004). However, the subject of our case study, Shenzhen, is partly made up of mountainous areas that could not be reached by traffic lines. Therefore, a road network, a contour with a 75-meter interval, and land use clusters are used to divide the land use space of the case study area. The obtained LUZs are covered by one and only one land use type.

With the use of such a division method, 5,181 LUZs are established in Shenzhen. According to local policies in Shenzhen, water bodies should not be reduced, and reserves and areas with an elevation that exceeds 80m are restricted from development. LUZs covered by water bodies, reserves, or with elevations that exceed 80m are referred to as water-body-constrained, reserve-constrained, and elevation-constrained LUZs, respectively. A total of 3,639 LUZs remain after eliminating these types of LUZs; hence, the optimization space and cost are reduced. Figure 11.2 shows the LUZs and the water-body-constrained, reserve-constrained, and elevation-constrained LUZs.

Assuming that raster data are used in this study, and taking 20m × 20m as the image resolution, 4,844 × 3,875 or 18770500 cells are needed in the grid to reflect the spatial land use allocation in Shenzhen, which is 18770500 / 5,181 = 3,622.95 times that of the vector data used in this publication. Using vector data obviously results in an efficient optimization system. However, vector data are irregularly organized; hence, spatial operations, such as judging the neighborhood for each polygon, are difficult.

11.2.2 Land Use Attribution Organization

A series of attributions is attached to vector LUZs. These land use attributions can be divided at the overall land use plan and the urban master plan levels. Such division is useful in optimization.

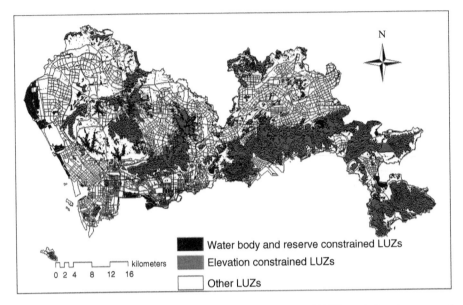

Figure 11.2 LUZs, water-body-constrained and reserve-constrained LUZs, and elevation-constrained LUZs

Municipal overall land use plan level

All 3,639 LUZs are optimized. Various attributions of each LUZ are first collected and organized to calculate the land use objectives. The spatial attributions include slope, elevation, soil type, vegetation coverage, precipitation, and assigned land use type of a given LUZ. In addition to these structured spatial attributions, the neighbors and their individual weight for each LUZ are treated as unstructured spatial attributions because the number of neighbors varies between the LUZs. All these attributions vary from one LUZ to another.

In addition to spatial attributions, other attributions are determined in accordance with the land use type. That is, LUZs with a given land use type maintain the same attribution values, such as, the NPS pollutant export coefficient, the GDP value generated per unit of land use area, and the carbon emissions factor from each land use type. These attributions are referred to as land use type-based attributions and vary from one land use type to the next.

Another attribution type is land use pair attribution, examples of which are the compatibility index and the changing cost. These attributions are determined by using a pair of land use types simultaneously instead of using only one land use type.

Urban master plan level

The urban extent of Shenzhen in the projected year can be determined by land use optimization. ULUZs located at the certified urban extent are optimized at the urban master land use plan level. Futian comprises 982 LUZs, which are developed as urban areas in accordance with the optimization at the last level. These 982 LUZs are converted into ULUZs and act as genes in the GA. The additional exhaustive

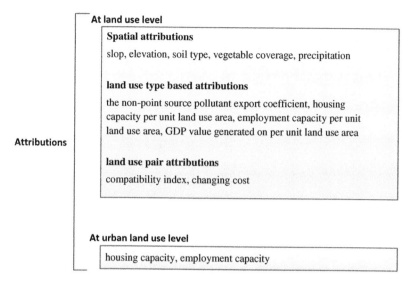

At land use level

Spatial attributions

slop, elevation, soil type, vegetable coverage, precipitation

land use type based attributions

the non-point source pollutant export coefficient, housing capacity per unit land use area, employment capacity per unit land use area, GDP value generated on per unit land use area

Attributions

land use pair attributions

compatibility index, changing cost

At urban land use level

housing capacity, employment capacity

Figure 11.3 Organization of attribution at land use and urban land use levels

attributions of these ULUZs are added at the urban master plan level; these attributions include the housing capacity and the employment capacity per unit of ULUZ. Figure 11.3 shows the branches of land use attributions at two levels. These spatial or non-spatial attributions are required in the MOO to evaluate the multiple objectives. Details of these spatial or non-spatial attributions will be given in Chapter 5 during the objective evaluation process.

11.2.3 Specification of the Optimization Process

Genetic algorithm

The two-level, GA-based MOO approach is used in this publication to solve multiple objective problems in the overall land use plan and in the urban land use plan. The specifications of the MOO approach are stated as follows (Feng & Lin, 1999; Cao *et al.*, 2012).

Step 1) Encoding and setting parameters. The possible land uses are assigned to the LUZs, and the population size and mutation probability of the GA should be set before the MOO starts. At the overall land use plan level, eight basic land use types comprise the possible land uses, whereas at the urban land use plan level, 33 urban land use types exist. Population size is assigned as 100 with a mutation probability of 0.05. Increasing population size increases the time and cost of computation. By contrast, a small population size easily leads to a local optimum through optimization. Based on existing studies (Balling *et al.*, 1999; Balling *et al.*, 2004), a population size of 100 is ideal to avoid huge time consumption and to reach local optimum.

Mutation probability denotes the probability of random changes in the genes. These random changes reduce the probability of the local optimum and may destroy the optimal results. Based on existing studies, a mutation probability of 0.05, which is appreciated by most scholars when performing land use optimization by using GA, is used in this publication.

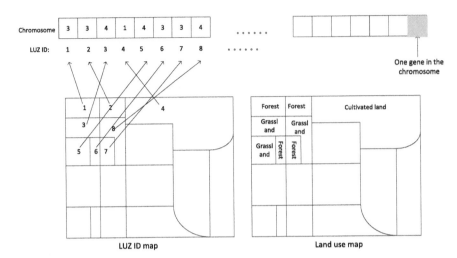

Figure 11.4 Relationship between the components in the GA and the actual spatial land use

Step 2) Generating the population in the first generation. The population is defined as a set of plans; 100 plans comprise a whole population. At each level, a chromosome with a number of genes equal to the number of LUZs or ULUZs in Shenzhen represents one plan. In this dissertation, one gene denotes one LUZ or one ULUZ, which is assigned an integer ranging from one to the maximum number of possible land use types. Each gene in a chromosome has an ID, which corresponds to the ID of the LUZ or ULUZ in 2D space. This ID is used to make the genes in the chromosome correspond to the LUZ or ULUZ in the GIS. The relationship between the components in the GA and the spatial land use in the GIS is illustrated in Figure 11.4.

The chromosomes in the first generation are generated randomly. At the overall land use plan level, each gene in the first generation is assigned a random integer ranging from 1 to 8; at the urban master land use plan level, each gene in the first generation is assigned an integer ranging from 1 to 33. Each integer denotes a specific land use or urban land use.

Step 3) Seeking initial feasible solutions. In this step, chromosomes that meet the constraints of the Shenzhen overall land use plan or urban land use plan are searched for across the whole population. The chromosomes that meet the constraints are considered feasible plans. Only the feasible plans are optimized in the following procedures.

Step 4) Calculating fitness. The fitness of each feasible chromosome is calculated. Firstly, the values of the objective of each plan or chromosome are computed. Then, the fitness of each plan can be determined based on the Maximin fitness calculating function. Fitness indicates the goodness of one plan or one chromosome. Sorting fitness allows the determination of the non-inferiority plan in one population.

Step 5) Selection. The best plans in one population are chosen as the parents for the next generation. Many methods can be used to select a suitable parent plan. Almost all methods involve the use of the ratio of the fitness of a certain plan to the summed fitness of the whole population to reflect the selected probability of that plan. As the population size in this dissertation is 100, the top 10% of the plans are most likely to be selected as parents. Therefore, the top 10 plans are selected, and the selected probability is determined by the value of fitness. A large fitness indicates a large probability.

Step 6) Crossover. With the parents selected in the prior procedure, the genes in the mother and father are exchanged. A binary chromosome is randomly generated with 0 and 1 as genes. The genes in the mother and father at the same allocation as the genes in the binary chromosome with a value of one are exchanged. The crossover can be carried out by using this binary chromosome. The new chromosome generated by the crossover is one population in the next generation. The crossover is followed by mutation.

Step 7) Mutation. For every gene that has been modified by the crossover process, a random number ranging from 0 to 1 is generated. A gene with a random number smaller than the mutation probability is randomly assigned another possible land use type. In this manner, the local optimum is easily avoided. Crossover and mutation continue until the number of offspring for the next generation reaches the population size, after which the process returns to Step 3 until the improvement of fitness between the last generation and its previous generation is small enough.

The GA process remains similar at the overall land use plan and the urban land use plan levels, but the possible land use types, specific objectives, and constraints are different. Optimization at the urban land use level is carried out up to the built-up land that is determined by optimization at the overall land use plan level.

Selection of Pareto solutions

After optimization by GA, the Pareto solutions among the whole solution pool provided by the GA should be selected. The methods of seeking Pareto solutions for multi-objective problems have been widely discussed (Benayoun *et al.*, 1971; Wierzbicki, 1980; Steuer & Choo, 1983; Huang *et al.*, 2008). This publication uses an adaptive algorithm proposed by Huang *et al.* (2008). The algorithm for minimizing objective includes the following steps:

Step 1) The minimum and maximum solution of each objective, Z_k, are found, and then the utopian point $p = (Z_1^*, Z_2^*, ..., Z_m^*)$ and the upper right point $g = (g_1, g_2, ..., g_m)$ are defined, both of which form the frame of the exploration domain. This box is added into the list of unexplored regions.

Step 2) The largest unexplored region R is deleted from the list if the list is not empty, and a new searching direction w is defined from R's upper right vertex g as $w = (1/g_1, 1/g_2, ..., 1/g_m)$.

Step 3) The following equation is solved:

$$
\begin{aligned}
&\text{Min } \beta \\
&w_k(Z_k(X)) \le \beta \ \forall k, \ 1 \le k \le m \\
&w_k > 0
\end{aligned}
\tag{11.1}
$$

where β is the minimum of such a problem, and m is the number of objectives.

Step 4) If the searched solution is already known, Step 2 is performed. Otherwise, a new solution is searched; then, the explored region is subdivided based on this point. For each sub-region, the coordinates of the upper right vertex and the unexplored volume are calculated. The list of unexplored regions is updated, and Step 2 is performed.

Two-level optimization

The optimization is carried out at two levels. Firstly, each LUZ in Shenzhen acts as a gene in the GA, and the optimization is carried out for all LUZs. Then, according to the results at the first level, the LUZs/genes which are assigned as built-up land, will be used in the second level. At the second level, the LUZs which are determined as built-up land are divided into urban land use zones. Usually, one LUZ will be divided into more than one ULUZ. Then, the optimization will be carried out at the second level. At this level, the land use types, objectives are different from those at the first level.

11.3 Results

11.3.1 Optimization at the Municipal Overall Land Use Level

Through means of optimization, the values of the maximizing objectives increased, the values of the minimizing objectives decreased, and the optimal Pareto solutions were achieved.

Improvement of objective value

Figure 11.5 and Figure 11.6 show the variations in the average objective value from the first generation to the last generation for the maximizing and minimizing objectives, respectively. Obviously, most maximizing objectives tended to increase, whereas most minimizing objectives tended to decrease from the first generation to the last. This result indicates that the MOO searched for the optimal solutions. However, the GDP value descended, and the NPS pollutant ascended; both variables did not act as expected. Some explanations of GDP and NPS pollutant are given below.

The GDP value of Shenzhen in 2008 was 7.79×10^{11} RMB according to the statistical yearbook (Shenzhen Statistics Bureau, 2009). Summing up the product of the unit GDP generated per unit area of a given land use type and the area of that given land use type, the GDP in 2008 was 7.79×10^{11} RMB, which is the same as that in the statistical data (see Table 11.1). Assuming that the land use structure in 2020 remains the same as that in 2008, denoting a scenario that is not optimized, and calculating the GDP by using the unit GDP value generated per unit area in 2020, the GDP value increased to approximately 1.77×10^{12} RMB. This increase results from the efficient land productivity in land use and is independent of the land use structure or spatial optimization. According to the optimization, the GDP was stable at approximately 2.00×10^{12} RMB (see Figure 11.4) and one Pareto solution is listed in Table 11.4 with a GDP of 2.01×10^{12} RMB as an example. A comparison between the GDP value in 2008 and one Pareto solution in 2020 indicates that the GDP value increases in accordance with the optimization. Even if the GDP decreases in the variations from the first to the last generation, the final GDP value is always larger than the GDP value in 2008. The decrease in the GDP value in the optimization process suggests competition between all of the objectives. However, the optimization of land use structure and space has already improved the GDP value.

The NPS pollutant in the first generation is less than 2,900 tons, which increased to 2,980 tons (see Figure 11.6). Referring to the land use pattern in 2008, the NPS

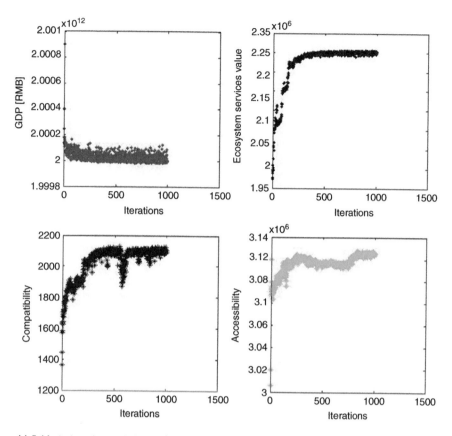

Figure 11.5 Variations in maximizing objective value from the first generation to the last generation

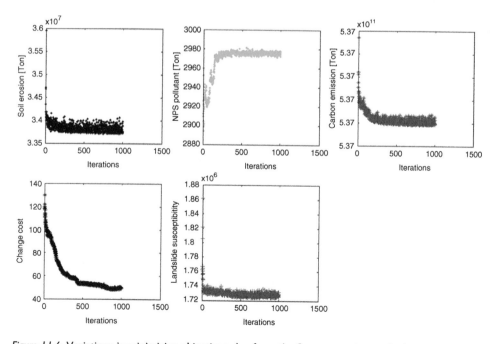

Figure 11.6 Variations in minimizing objective value from the first generation to the last generation

Table 11.1 GDP comparison

Land use type	GDP generated on unit land use area		Area in 2008 [sq. km]	Area in 2020 (one Pareto solution) [sq.km]	GDP in 2020 (one Pareto solution) [RMB]	GDP in 2008 [RMB]
	In 2020	In 2008				
Cultivated land	39.5336	35.38	32.06	25.51	1008351610	1134282800
Garden plot	0.526451	0.08	231.46	94.43	49710502.72	18516800
Forest	0.100731	0.09	594.41	638.65	64331929.47	53496900
Grassland	0	0	2.61	0.69	0	0
Built-up land	2,339.72	1,031.04	754.43	859.90	2.01193E + 12	7.77848×10^{11}
Transportation	0	0	187.38	187.38	0	0
Water body	1.059306	0.13	141.34	141.34	149722310	18374200
Unused land	0	0	9.14	4.95	0	0
Sum	–	–	–		2.01×10^{12}	7.79×10^{11}

pollutant is 2,907.90 tons, which suggests that the NPS pollutant in the first generation is impractical because the system considers other objectives that may conflict with the minimizing objective of the NPS pollutant. The NPS pollutant value in the last generation is around 2,980 tons, whereas the NPS pollutant reaches 3,087.33 tons in the plan proposed by the Shenzhen government (see Table 11.3). This difference indicates that optimization minimizes the NPS pollutant and compromises all of the conflicting objectives.

Variations were large during the first 200 iterations (Figure 11.5 & Figure 11.6). For example, landslide susceptibility decreased from 1.86×10^6 to about 1.74×10^6 in the first 200 iterations, whereas the variations were obscure in the subsequent 800 iterations. The same pattern was observed in other objectives. At first, a large space in which to perform the optimization was available because the land use plans were initially generated at random. As the optimization process continued, land use allocation became optimized. In such a case, improving land use allocation that has already been optimized is difficult. At the same time, objectives that competed with one another aggravated the difficulty of optimization. Therefore, the improvements became minimal and exceptionally difficult to achieve when the optimization continued. Variations in the objective values were motivated by changes in land use structure and spatial distribution of the total land use in Shenzhen.

Land use structure

Figure 11.7 shows the land use structures before and after optimization along with the constraints on the areas of specific land usages, where Opt_mean denotes the average area of all the optimal plans generated by the MOO, Opt_min denotes the minimum value of all optimal plans, Opt_max denotes the maximum value of all optimal plans, and Con_low_limit and Con_up_limit denote the minimum and maximum limitations of a certain constraint, respectively. The graph in Figure 11.7 suggests that the optimal land use structure met all the constraints. Most constraints on the land use structure were proposed by local policies or national laws. Thus, satisfying these constraints

Table 11.2 Analysis of land use structure constraints

Land use type	Min	Max	Mean	Current land use pattern	Municipal overall land use plan by Shenzhen government	Constraints
Cultivated land	2,548.627	3,973.827	2,613.15	3,206	6,197.16	> 2,465
Garden plot	9,442.57	9,442.57	9,442.57	23,146	28,160.47	–
Forest	6,2376.14	63,879.02	63,735.25	59,441	59,144.27	–
Grassland	68.72032	68.72032	68.72032	261	24.98	–
Built-up land	85,978.08	86,656.32	86,057.33	75,443	71,382.74	> 85,800 and < 109,133
Transportation	18,738	18,738	18,738	18,738	20,880.33	–
Water body	14,134	14,134	14,134	14,134	5,786.55	= 14,134
Unused land	494.99	494.99	494.99	914	3,707.51	–

made the optimal results easy to apply in practice. Table 11.2 lists the specific numbers shown in Figure 11.7.

With regard to the Opt_mean, the built-up land and forest land increased, whereas the cultivated land and garden plot decreased. The increase of built-up land is required for population growth and economic development; the increase of forest land is required to address environmental issues, such as carbon emissions, NPS pollution, soil erosion, and ecosystem service value. Forest land has a higher environmental protection value than cultivated land and garden plots. Given that the total area of Shenzhen is constant, certain land use types increased while others decreased. Cultivated land and garden plots cannot provide as much environmental benefit as forest land. Thus, these two land uses were reduced in the MOO process. Notably,

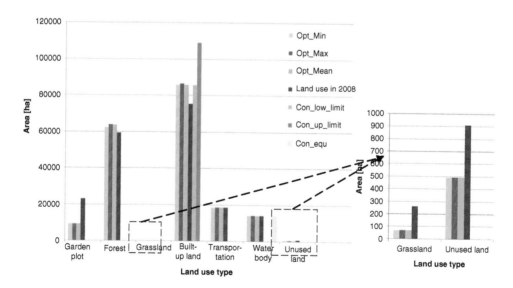

Figure 11.7 Comparison of land use structure in terms of optimal results and corresponding constraints

conversion of cultivated land is controlled by law; therefore, continued reduction is impracticable. As a result, the area of optimal cultivated land is slightly larger than the constraint of 2,465 hectares (see Table 11.2).

The land use structure of the municipal overall land use plan proposed by the Shenzhen government is listed in Table 11.2. Even if the plan allows for enough cultivated land, the built-up land is much smaller than the built-up land in 2008 and may not be enough to accommodate the increasing population in Shenzhen. The body of water is decreased by the plan, whereas the forest land remained constant. In the optimal land use plans, the water body did not decrease, while the forest land increased, thereby leading to an environment-friendly land use pattern.

In summary, the optimal land use structure generated by the MOO is more environment-friendly and provides more built-up land than the land use structure specified in the government plan.

Pareto solutions

Although the solutions of the last generation were obtained by using MOO, decision-makers need to select from the Pareto solutions in the last generation instead of all the solutions. This section discusses the Pareto solutions. Nine objectives were identified in the land use optimization system. The results of the Pareto solution were obtained by considering all the objectives. However, illustrating the Pareto solutions visually in a nine-dimensional space is impossible. In this case, the objectives were compared in pairs to illustrate the Pareto solutions, and there are Pareto solutions for nine objectives rather than two objectives.

Figure 11.8 and Figure 11.9 illustrate the variations of four maximizing objectives in pairs: GDP value; ecosystem service value; compatibility; and accessibility. The red points in Figure 11.8 and Figure 11.9, which are solutions in the last generation, are

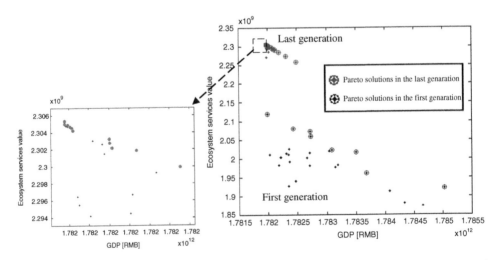

Figure 11.8 Pareto solutions for the first and last generations, with GDP value as the x-axis and ecosystem service value as the y-axis

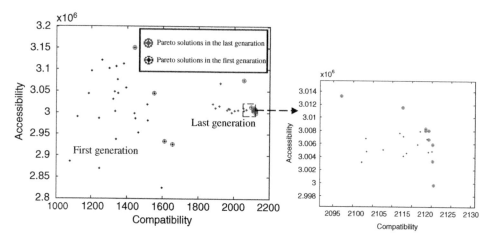

Figure 11.9 Pareto solutions for the first and last generations, with compatibility value as the x-axis and accessibility as the y-axis

farther away from the origin than the blue points, which are the solutions in the first generation. These results indicate that both maximizing objectives were higher than their values in the first generation.

In addition to this increase in value, the distributions of the scatters were different. The blue points in the figures are distributed in a dispersed manner, whereas the red points are distributed in a clustered manner, which indicates that the solutions became convergent through optimization.

The Pareto solutions in each generation were marked. No best solution was found when considering two objectives. The Pareto solutions improved significantly from the first generation to the last. Decision-makers can then select one plan from the Pareto solution pool.

The variations in the minimizing objectives from the first to the last generation are shown in Figure 11.10 to Figure 11.12. Unlike the maximizing objectives, the minimizing objectives neared the origin as the optimization continued. In Figure 11.10 to Figure 11.12, the maximizing objectives are represented as red points that are closer to the origin and are more clustered than the blue points, which are distributed in a dispersed manner. The Pareto solutions were marked.

To analyze the Pareto solutions in a detailed manner, the objective value of the Pareto solutions in the last generation was analyzed quantitatively. According to the definition of the Pareto optimal solution, 33 optimal Pareto solutions were obtained in the last generation. The minimum, maximum, mean, and standard deviations of the objective value of the 33 optimal Pareto solutions are listed in Table 11.3. Firstly, the gap between the maximum value and the minimum value was small because the competition between the objectives increased as the optimization progressed. Improving the objective value was difficult. Therefore, the differences between the Pareto solutions in the last generation were not significant. Secondly, the economic benefit improved greatly compared with the objective value of the land use pattern in 2008. However, environmental issues, such as soil erosion, NPS pollution, carbon emissions, and

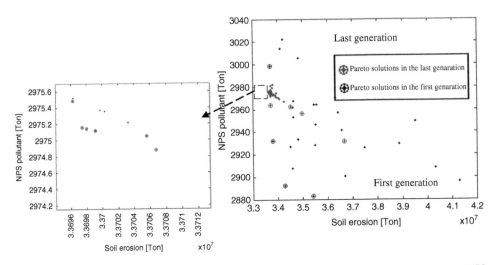

Figure 11.10 Pareto solutions for the first and last generations, with soil erosion as the x-axis and NPS pollutant load as the y-axis

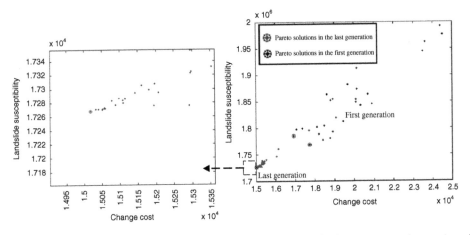

Figure 11.11 Pareto solutions for the first and last generations, with change cost as the x-axis and landslide susceptibility as the y-axis

landslide susceptibility, worsened, according to the objective values. A question that arises is why the optimization process led to poor land use conditions. The environment inevitably degrades in the urbanization process along with the growth of built-up land. In Shenzhen, the spatial planning of land use should first maintain economic growth and accommodate the increasing population. Under this constraint, the MOO attempted to properly allocate land use to maximize environmental preservation. Land use that is allocated without optimization is expected to degrade the environment to a greater degree than the conditions generated by the MOO. Meanwhile, the gap between the land use pattern in 2008 and the Pareto solutions was acceptable in terms of soil erosion, NPS pollution, carbon emissions, and landslide susceptibility. For

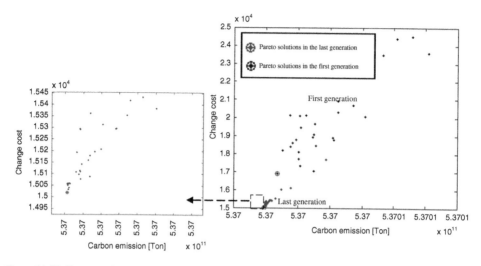

Figure 11.12 Pareto solutions for the first and last generations, with carbon emission as the x-axis and change cost as the y-axis

example, the soil erosion was 3.25×10^7 tons in 2008, which increased to 3.37×10^7 tons in the Pareto solution.

The objective values of the overall plan specified by the government are listed in Table 11.3 for comparison purposes. The government plan would generate less GDP

Table 11.3 Objective value analysis of Pareto solutions

Objectives	Characteristic	Max value	Min value	Mean	Standard deviation	Current land use pattern	Municipal overall land use plan by Shenzhen government
Economic benefic	To maximize	2.01×10^{12}	2.02×10^{12}	2.01×10^{12}	2.25×10^8	7.79×10^{11}	1.67×10^{12}
Ecosystem services value	To maximize	2.31×10^9	2.19×10^9	2.29×10^9	2.08×10^7	2.24×10^9	1.95×10^9
Soil erosion	To minimize	3.67×10^7	3.37×10^7	3.39×10^7	5.61×10^5	3.25×10^7	2.41×10^7
Non-point source pollution	To minimize	2,982.03	2,931.78	2,972.22	8.79	2,907.90	3,087.33
Carbon emission	To minimize	5.37×10^{11}	5.37×10^{11}	5.37×10^{11}	7.21×10^5	5.37×10^9	1.04E+12
Compatibility	To maximize	2,120.60	1,880.65	2,058.40	77.12	2,156.57	1,507.06
Change cost	To minimize	17,697	15,018	15,308	500	0	48,199
Accessibility	To maximize	3.07×10^6	3.00×10^6	3.01×10^6	1.61×10^4	2.70×10^6	2.84×10^6
Landslide susceptibility	To minimize	1.80×10^6	1.73×10^6	1.73×10^6	13,328.06	1.48×10^6	1.47×10^6

value because the built-up land in the government plan is less than that in the optimal plans; built-up land significantly contributes to GDP increase. However, having less built-up land in the future results in less soil erosion during construction. Moreover, given that the landslide susceptibility reflects the possibility of landslides on built-up and cultivated lands, where more human activities are carried out, having less built-up land leads to a lower landslide susceptibility. Therefore, the government plan is better than the optimal plans in terms of soil erosion and landslide susceptibility. However, in terms of the other objectives related to land use structure and spatial land use allocation, the optimal plans are better than the government plan.

A total of 33 Pareto optimal plans were illustrated by using radar graphs in groups of minimizing and maximizing objective plans; all objectives were first normalized (see Figure 11.13 & Figure 11.14). The radar graph shows that all the objectives are in competition with one another. For example, for the minimizing objectives, even if the objective values of soil erosion, landslide susceptibility, change cost, and carbon emissions approach 1, the NPS pollution in the plan in purple is the lowest among all the Pareto plans.

In addition to the statistical analysis, spatial distributions of two Pareto plans are shown in Figure 11.15. These two Pareto plans, which maintain the largest and smallest areas of built-up land, are compared in Figure 11.15. The built-up land in 2008 and the increased built-up land according to the Pareto plans are mapped. This mapping reveals

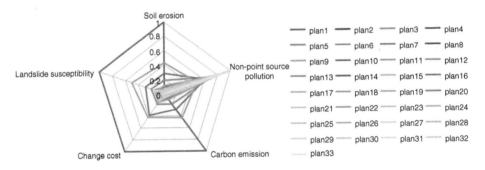

Figure 11.13 Radar graph with minimizing objectives for Pareto solutions

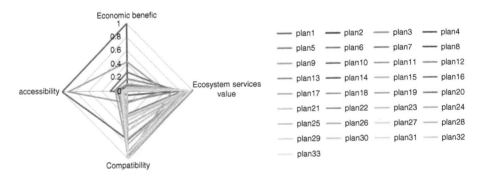

Figure 11.14 Radar graph with maximizing objectives for Pareto solutions

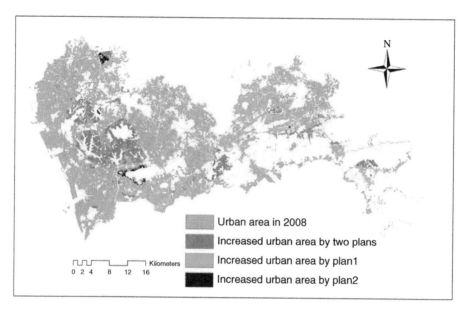

Figure 11.15 Spatial distribution of additional built-up land in the Pareto plans

that most of the increased built-up land generated by the Pareto plans overlapped (see the purple plots in Figure 11.15) and the second plan generates slightly more built-up land than the first plan. Hence, the final alteration between the Pareto solutions was small through the MOO.

The spatial land use pattern of one Pareto solution is shown in Figure 11.16. The spatial land use pattern in 2008 and the land use plan proposed by the government are shown in Figure 11.17 and Figure 11.18, respectively. According to the spatial distribution, the water body is maintained and acts as a constraint in MOO. Secondly, the outline of the total land use pattern in Shenzhen is maintained, with some of the cultivated land, garden plot, and grassland being converted into built-up land, thereby creating a compatible land use pattern. For example, the cultivated land and garden plot in the black cycle in 2008 (see Figure 11.17) disappears in 2020, which results in a cluster spatial distribution of built-up land. As specified in the government plan, less built-up land is generated, and the differences between the government plan and the land use plan in 2008 are small.

11.3.2 Optimization at the Urban Master Plan Level

In this study, Futian was used as an example in the spatial planning of land use at the urban master plan level. After the land use optimization of the whole of Shenzhen at the municipal overall land use level, the urban area of Shenzhen in the projected year was determined, and the optimization at the urban master plan level was conducted within the extent of the determined urban area. The planners or decision-makers should select from the Pareto plans because different Pareto plans provide various extents of an urban area. To simplify the process, one Pareto plan was randomly selected to provide

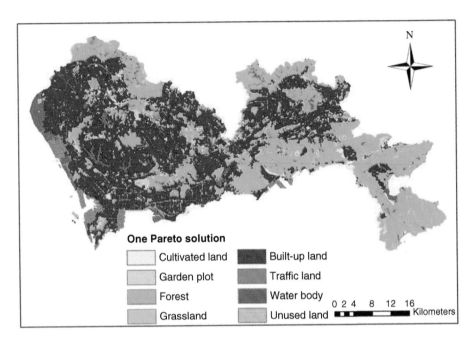

Figure 11.16 Spatial land use pattern of one Pareto solution

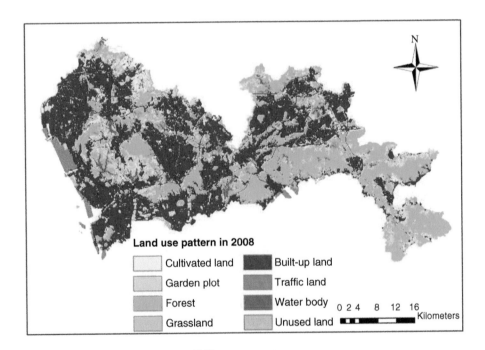

Figure 11.17 Spatial land use pattern in 2008

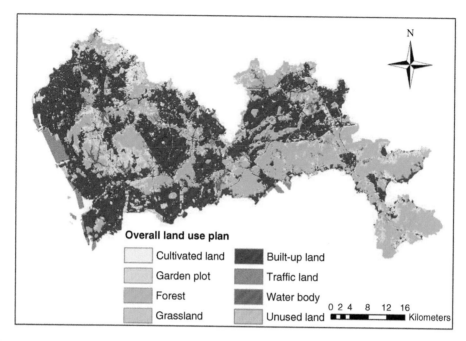

Figure 11.18 Overall land use plan outlined by the Shenzhen government

the extent of the urban area in the projected year. Within this context, the optimization for the spatial planning of land use at the urban master plan level could be conducted.

Improvement of the objective value

Similar to the results of the MOO at the municipal overall land use plan level, the maximizing objectives increased, whereas the minimizing objectives generally decreased (see Figure 11.19 & Figure 11.20). This result suggests that the MOO is effective at this level. For the minimizing objectives (see Figure 11.19), the changing cost and pollution from industrial land decreased from the first generation to the last generation; however, the Schutz index, which is used to measure equity, showed unexpected variation. The Schutz index sharply increased from approximately 139 to 144 in the first 100 iterations. Afterwards, the index decreased and returned to a value of approximately 140. Although the Schutz increased at first, it decreased thereafter. The variation in equity reflects the competition between the objectives. In the land use system, simultaneously improving all of the objectives while avoiding deterioration of the other objectives is impossible.

All maximizing objectives, except for accessibility and mixed land use, increased upon optimization (see Figure 11.20). The results indicated that the proposed approach effectively optimizes the spatial planning of land use at the urban master plan level. For example, compatibility, increased in the first 100 iterations; however, increasing compatibility became difficult after the 200th iteration. At first, the plans were randomly set far from the optimal condition. As optimization continued, the solution approached

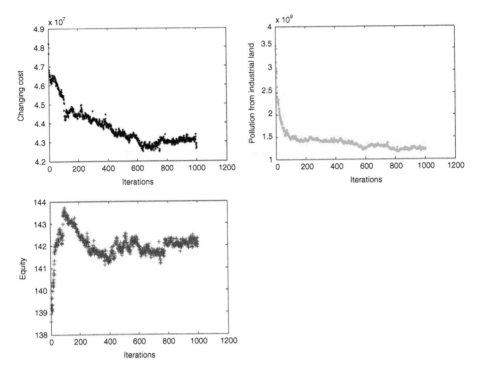

Figure 11.19 Variations in the values of the minimizing objective from the first generation to the last generation

the optimal condition, thereby causing difficulties in improving the specific objectives. Another mixed land use model was implemented. The value of mixed land use was set randomly at first because the MOO starts with a set of random plans. The value largely decreased before the first 500 iterations, but floated after the 500th iteration. During this floating process, a competition between the mixed land use and other land use objectives was observed.

Urban land use structure

The number of possible land use types on Futian's land use map is 33. Three land use types are not urban land use types but are allocated within the administrative scope of Futian. These land use types are nature reserves, water body, and forest lands. In the optimization, these non-urban LUZs were not considered at the urban master plan level. Therefore, the urban land use structure generated by the MOO did not include these three land use types, although they existed in the 2008 land use map of Futian.

Table 11.4 lists the mean, minimum, and maximum areas of each urban land use type in the Pareto solutions. Firstly, the urban land use structure met the requirements. The ratio of residential land was (2% + 10% + 9% + 2% + 1% + 1%) = 25%, the ratio of industrial land was (9% + 6%) = 15%, the ratio of green space was (5% + 6%) = 11%, and the ratio of public management and public service land was (2% + 1% + 1% + 1% + 1% + 1%) = 7%, all of which satisfied the land use structure requirements.

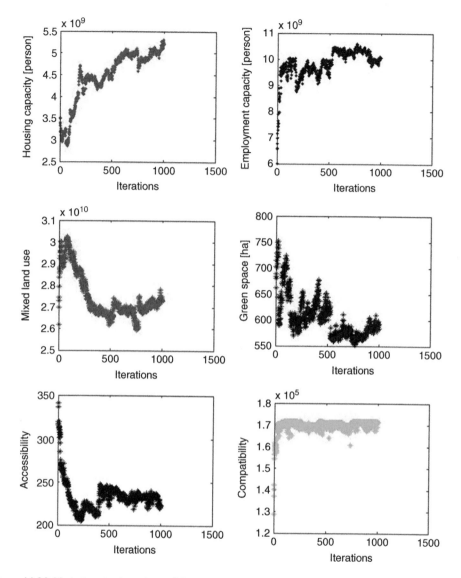

Figure 11.20 Variations in the values of the maximizing objective from the first generation to the last generation

The mean, maximum, and minimum values in this study refer to the Pareto solutions generated in the last generation. According to the optimal land use structure listed in Table 11.4, the ratio of the development of backup land (E9) is all zeroes. Backup lands are developed into built-up lands by using the MOO approach because the cost of developing backup lands is low and because their environmental contribution is likewise relatively low. This reason causes its ratio of zeroes in the final optimal plans, and this suggests that the MOO is effective at the urban master plan level.

Table 11.4 Urban land use structure

ID	A.b.	Urban land use in 2008			Pareto solutions		
		Area	Percentage	Mean area	Percentage	Max area	Min area
1	C1	153.09	2.8%	143.76	3%	129.50	155.04
2	C2	63.39	1.2%	209.65	4%	176.77	228.87
3	C3	54.93	1.0%	200.28	4%	172.29	222.08
4	C4	53.86	1.0%	189.98	3%	166.98	223.90
5	C5	110.56	2.1%	177.30	3%	161.42	203.94
6	C9	316.36	5.9%	278.66	5%	214.89	295.96
7	D3	162.43	3.0%	165.00	3%	119.15	190.70
8	D6	0.00	0.0%	0.00	0%	0.00	0.00
9	E1	0.00	0.0%	0.00	0%	0.00	0.00
10	E5	0.00	0.0%	0.00	0%	0.00	0.00
11	E9	309.44	5.7%	0.00	0%	0.00	0.00
12	G1	516.38	9.6%	270.91	5%	247.72	300.21
13	G2	433.67	8.0%	313.27	6%	286.33	331.86
14	IC1	18.51	0.3%	100.26	2%	75.50	119.56
15	IC2	106.96	2.0%	55.15	1%	41.91	65.36
16	IC3	82.15	1.5%	50.64	1%	36.75	63.54
17	IC4	62.06	1.2%	74.52	1%	61.74	93.31
18	IC5	53.40	1.0%	76.41	1%	67.47	90.57
19	IC8	90.60	1.7%	68.47	1%	53.17	83.61
20	M1	177.87	3.3%	497.61	9%	478.40	519.43
21	M2	39.10	0.7%	330.35	6%	313.70	347.86
22	R1	7.28	0.1%	85.22	2%	74.47	97.37
23	R2	1,906.83	35.4%	557.46	10%	521.54	616.08
24	R3	207.55	3.9%	491.05	9%	430.23	545.74
25	R6	191.45	3.6%	109.81	2%	72.88	150.68
26	R7	8.31	0.2%	72.21	1%	59.95	85.63
27	R8	31.09	0.6%	73.58	1%	60.98	103.93
28	W1	9.56	0.2%	190.54	3%	162.62	218.00
29	S	89.13	1.7%	135.01	2%	97.70	195.08
30	U1	66.10	1.2%	206.93	4%	180.11	226.30
31	U3	4.72	0.1%	120.34	2%	107.89	152.49
32	U4	41.44	0.8%	124.72	2%	91.82	154.74
33	U5	21.71	0.4%	83.16	2%	62.28	105.95

In addition, the ratios of the two most important land use types, namely, residential and commercial lands, changed. In 2008, the ratio of residential land was relatively high; that is, the ratio of second-class residential land in 2008 was 35.4%, whereas the total ratio of residential land in 2020 has been projected to reach 25%. The ratio of commercial land in 2008 was 14% and it is projected to reach 22% by 2020. Residential land provides housing for the citizens, whereas commercial land provides employment for the citizens. An increase in commercial lands leads to economic development. However, restructuring first-class, second-class, and third-class residential lands can increase housing capacities even if the total ratio of residential land decreases. In 2008, the ratio of third-class residential land was 3.9% and it is projected to increase to 9% by 2020. Third-class residential land provides an intensive land use mode that offers a 520 housing capacity per hectare, whereas first-class and second-class residential lands offer 100 and 463.37 housing capacities per hectare, respectively. Thus, the MOO aimed to maintain economic growth in Futian while providing citizens with adequate housing.

Pareto solutions

At the urban master plan level, optimization was conducted for land uses in the urban area. Similar to the results at the municipal overall land use plan level, the Pareto solutions were obtained by considering nine objectives; however, illustrating them in a nine-dimensional map is difficult. Therefore, the Pareto solutions are illustrated in pairs. The results indicate that 15 plans satisfied the constraints and were considered feasible. A total of 15 Pareto plans were established in accordance with the Pareto solution searching approach. In this phase, the Pareto solutions were analyzed.

Figure 11.21 and Figure 11.22 illustrate the Pareto plans of minimizing objectives in pairs. The objectives were optimized, that is, the solutions for the last generation moved closer to the origin compared with the solutions for the first generation. These results suggest that the MOO is effective at this level. Moreover, several Pareto solutions exist when two objectives are considered. Thus, only a series of non-inferior solutions are available even in a 2D optimization space.

The Pareto solutions for maximizing objectives are represented in pairs in Figure 11.24 to Figure 11.25. In Figure 11.23 and Figure 11.25, the solution points are far from the origin, which thus indicates an increase in the objective value from the first

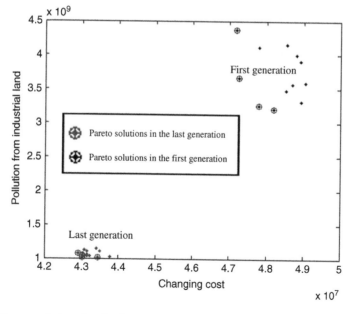

Figure 11.21 Pareto solutions for the first and last generations, with changing cost as the x-axis and pollution from industrial land as the y-axis

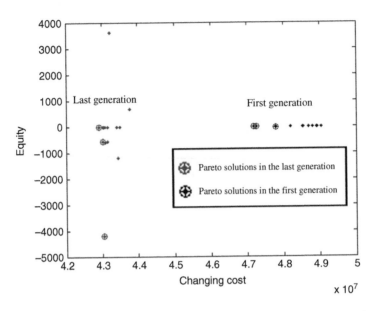

Figure 11.22 Pareto solutions for the first and last generations, with changing cost as the x-axis and equity as the y-axis

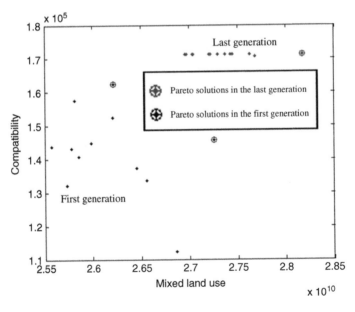

Figure 11.23 Pareto solutions for the first and last generations, with mixed land use as the x-axis and compatibility as the y-axis

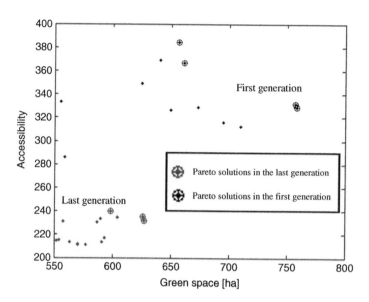

Figure 11.24 Pareto solutions for the first and last generations, with green space as the x-axis and accessibility as the y-axis

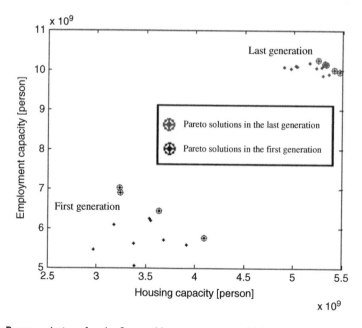

Figure 11.25 Pareto solutions for the first and last generations, with housing capacity as the x-axis and employment capacity as the y-axis

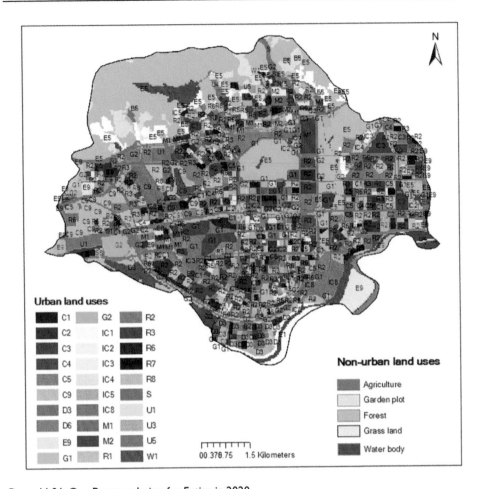

Figure 11.26 One Pareto solution for Futian in 2020

generation to the last generation. However, in Figure 11.24, the maximizing objectives of green space and accessibility are close to the origin as a result of the optimization. Superficially, this result suggests that the MOO is unprofitable for these two objectives. However, taking the whole objective system into account, the competition between all of the objectives decreases the maximizing objectives.

To directly understand the Pareto solutions, the spatial distribution of one Pareto solution is shown in Figure 11.26. Compared with the urban land use pattern in 2008, the optimal land use pattern offers more mixed spatial land use distribution. Various land use patterns lead to more social equity in terms of spatial land use.

11.4 Conclusion

Shenzhen has seemingly unresolvable land scarcity and land use planning problems. Therefore, the spatial planning of land use for Shenzhen at two levels was optimized to guide the overall land use and urban master plans.

At the first level, the spatial planning of overall land use was conducted. The overall land use objectives were proposed based on the characteristics of Shenzhen, the land use planning principles of sustainable land use, smart growth and new urbanism, and the policies and laws of China. At this level, the land use structure and the spatial allocation of each land use were optimized. Most importantly, the extent of the urban area was determined considering the multiple objectives and constraints, which were then used as the scope for the spatial planning of land use at the urban master plan level. At the overall land use planning level, the objectives referred to issues that affect the entire city, such as cultivated land protection, elevation constraints, water body protection, built-up land growth to accommodate development, and other environmental problems. The overall land use structure and spatial allocation could be determined while considering all the proposed objectives and constraints by using the MOO approach.

Based on the results of the first-level optimization, spatial planning of land use at the urban master plan level was conducted. This level is concerned with urban land use allocation within the urban areas and not across the entire city. The defined objectives reflect the urban development goals, such as housing capacity, employment capacity, green space, equality in urban areas, and other urban issues. The impossible urban land uses were allocated to the spatial space of the urban area by using the MOO.

With the use of this two-level MOO approach, the goals at different land use levels were carefully defined, thereby achieving reasonable results. According to Chinese policies, the urban land use plan should be implemented based on the results of the overall land use plan. Inconsistency between the overall and the urban land use plans results in difficulties in implementation. From a technical perspective rather than from a policy perspective, this study proposes an approach where the overall and the urban land use plans are connected in one system.

In the process of defining the objectives and constraints, statistical data on population growth, built-up land increase, economic development, and local and national policies were used to calculate some of the coefficients. For example, the unit employment capacity for people per hectare was defined by the land use data and the corresponding jobs provided by each land usage in history. Traditionally, given the difficulty in data collection and other related issues, the unit employment capacity in people per hectare and the unit housing capacity per hectare were determined based on the literature review and the suggestions of the planners and decision-makers. This setup causes the entire system to become stochastic and produces unreasonable objective and constraint calculations. Moreover, the total number of built-up lands is an extraordinarily significant index that determines economic development and population growth, and significantly affects the environmental system. To set the constraint to the built-up land area, the population forecasting model, the per capital built-up land forecasting model, and the rules from Shenzhen's local and national policies were considered in one system. The constraint on the area of built-up land was assigned by coordinating the technique model and the policies.

Spatial problems in spatial planning are more complex than the coefficient approaches used to evaluate non-spatial problems. Therefore, some innovative methods are used to increase the accuracy of the results. In this publication, although measuring the spatial relationship of vector data is difficult, vector data were used in the optimization to decrease the amount of data produced and to improve the efficiency of the optimization.

Within this context, the spatial weighting matrix, which measures the spatial relationship between neighborhoods, was used to define the effect of one neighbor on the center LUZ. The spatial weighting matrix ranged from 0 to 1, which suggests that the effect of the neighbor on the center LUZ changed. The traditionally used binary index merely reflects whether or not two LUZs are neighbors. Spatial analysis was used to evaluate and measure the spatial objectives. Among all the objectives, spatial landslide susceptibility, spatial distribution of NPS pollution, soil erosion, and spatial distribution equity were spatially related. Taking spatial equity as an example, the global Moran's I index was used to measure the spatial autocorrelation.

In summary, the approach outlined in this publication is a multiple-objective approach for the spatial planning of land use at two levels. This approach can be applied to balance economic, social, and environmental objectives and constraints in a land use system. The proposed approach offers a means to evaluate and measure the multiple concerns and criteria of the usually conflicting and competing objectives within a land use planning system.

Part IV

Summary and Contributions

Chapter 12

Summary and Contributions

In this chapter, the summary of the whole book is presented. In Part One, the causal factors driving land use changes and the basic situation of the case study were reviewed. Then, the excellent performance of three novel models, namely, GTWLM, ST-PLM, and GSTLM, were described, respectively. Finally, the findings of the spatio-temporal analysis of land use change are detailed.

Part Two reviewed the land use multi-objective optimization which is proposed herein. It firstly concludes the needs of distinguishing two optimization levels, and presents objectives and constraints, respectively. Then, the process of selecting the Pareto solutions is summarized. At the end, this book lists some of the contributions to MOO study field.

12.1 Spatio-temporal Analysis Method

The study of land use change herein is a prerequisite to understanding the complexity of land use change, forecasting the future trends of land use change, and evaluating the ecological impacts of land use change. In order to better address some important issues in land use change study, this research aims to construct novel statistical models to establish spatio-temporal relationships between land use and explanatory factors. The outcome of this research, which focuses on the analysis, could benefit urban planners and policy-makers in their efforts to effectively understand the land use change process from a spatio-temporal perspective.

A review of the causal factors driving land use changes reported in the literature is provided in this study. Furthermore, a variety of techniques for land use change modeling are briefly reviewed and discussed. In particular, studies of spatio-temporal models of land use change are introduced. The advantages and limitations of these techniques are also provided. It is found that some challenges still exist within the domain of 'spatio-temporal modeling' for land use change. With due consideration to the powerful explanatory capacity, the logistic regression framework is selected in this research to address three major problems: spatio-temporal non-stationarity; spatio-temporal autocorrelation; and individual effect.

Ever since China's economic reform, the city of Shenzhen has undergone dramatic growth and development. During the past decade, Shenzhen has witnessed drastic changes in its land use change pattern. Considering its 'one city, two system' framework, its SEZ was selected as the study area for this research. Because the data involved in the study area were very plentiful, a random sampling from different years was performed to obtain the dataset. In view of the aforementioned problems, this research proposes three spatio-temporal logit models for studying the land use change. These are the GTWLM, the ST-PLM, and the GSTLM, which have the potential to effectively address the research challenges.

The GTWLM, which is able to consider spatio-temporal non-stationarity, includes temporal data in a spatio-temporal framework by proposing a spatio-temporal distance. Also, TWLM and GWLM, which consider temporal non-stationarity and spatial non-stationarity, respectively, are introduced. Compared with the global model (MNLM), TWLM and GWLM increase the PCP values from 74.1% to 75.4% and 79.2%, respectively, and GTWLM yields a considerably higher PCP of 82.3%. McNamara's test shows that the differences between those models are significant. The Kappa coefficients also reveal that the GTWLM is better than MNLM, TWLM, and GWLM. More importantly, TWLM, GWLM, and GTWLM can provide much more information on spatial and temporal relationship, which facilitates model development and leads to a better understanding of the land use change processes.

In the ST-PLM, spatio-temporal autocorrelation is considered in the random individual effect component ε_i^j with an assumption that the autocorrelation between ε_i^j is inversely proportional to the spatio-temporal distance between them. The ST-PLM incorporates the spatio-temporal autocorrelation and individual effect in one model, which is ignored by the MNLM. As a result, the ST-PLM model achieves a higher overall PCP (79.4%) than the MLNM (74.1%). Also, the accuracies for undeveloped, industrial, and commercial/transporation/others achieved by the ST-PLM are better (i.e., 86.4% vs 90.4%; 50.7% vs 56.0%; and 64.1% vs 76.6%). Both the McNamara's test and the AIC test corroborate the superior performance of the ST-PLM. Besides, the Kno evinces that the ST-PLM consistently achieves a better result than the MNLM. Specifically, the ST-PLM's ability to specify location is better than the MNLM's, and the two models have the same ability to specify quantity.

Considering all of the challenges in one model is an attempt to concentrate on the integration of GTWLM and ST-PLM. Thus, the GSTLM has been proposed to explore spatio-temporal non-stationarity and autocorrelations simultaneously, whilst considering the individual effect. The experimental results of the case study in the SEZ demonstrate that the GSTLM achieves a better modeling accuracy than both the GTWLM, which deals with spatio-temporal non-stationarity only, and the ST-PLM, which deals with spatio-temporal autocorrelation and individual effect only. Compared with the other models, the GSTLM can improve the PCP of MNLM, GTWLM, and ST-PLM from 74.1%, 82.3%, and 79.4% to 85.9%, respectively. The McNamara's test shows that there is a significant difference between MNLM, GTWLM, ST-PLM, and GSTLM. Here, the PCP improvement of GSTLM on MNLM is less than the sum total of the improvement on MNLM by GTWLM and ST-PLM. This is so because allowing for non-stationarity in the regression parameters can account for at least some, and possibly a large part of the autocorrelation in terms of error in a global model calibrated with spatio-temporal data. The Kappa coefficients also indicate that the GSTLM is the most successful model. The results of the t-test also demonstrate that the individual effect is significant in most of the observations and that spatio-temporal autocorrelation exists in some observations. Similar to the GTWLM, the GSTLM also allows the model parameters to vary across space and time, which provides deep insights into the spatio-temporal variations of the land use pattern. It has been shown that the spatio-temporal variability of each factor which influences the land use pattern leads to different patterns.

Overall, this book addressing the spatio-temporal analysis of land use change has led to the following findings:

1. Individual effect, non-stationarity, and autocorrelation in space and time do exist in this study area. The proposed models, which consider those problems, can greatly improve the accuracy and reliability of land use change study especially in a spatio-temporal framework.
2. The form of spatio-temporal distance, which is obtained by combining the spatial distance and temporal distance, is case-based. For land use change study in the SEZ, Shenzhen, the balance ratio between spatial distance effect and temporal distance effect (τ) is 0.1826.
3. The small value of τ suggests that the spatial effect is greater than the temporal effect in this study area. The main reason is that only seven temporal periods are available. The inadequacy in temporal information can be expected to degrade the model performance of TWLM which considers temporal non-stationarity only.
4. The selection of a particular spatio-temporal weighting function significantly influences the results. In this study, the exponential function with an adaptive bandwidth was adopted.

The parameter estimates demonstrate a remarkable spatial and temporal change in SEZ, Shenzhen. Most of the parameters show a significant difference between Yantian and the other districts. It indicates that the determinants of the land use pattern have varying effects in different locations. Although the temporal space of the study is only ten years, the temporal variation of parameter estimates still exists.

12.2 Land Use Multi-objective Optimization

The study in this book proposes an MOO-based spatial planning of land use in the Chinese context. The spatial planning of land use focuses on the spatial layout of land use and therefore it leads to a beneficial effect on the implementation of land use planning in space. In the Chinese context, the spatial inconsistencies between the municipal overall land use plan and the urban master plan have been the subject of debate for years. This study attempts to coordinate these two plans in space by using the spatial planning of land use. Using a computer to detect, evaluate, and measure these objectives is necessary because multiple objectives are involved in the land use planning system, and these objectives usually compete with one another. Making the situation increasingly difficult is usually the nonlinear and unstructured nature of spatially related objectives, posing measurement difficulties for the planners. Therefore, the MOO approach is used in this study to induce a trade-off between all these objectives. The GIS, RS, and statistical models are used to evaluate and measure the land use objectives.

The objectives proposed in the MOO significantly affect the final optimized plans. In this study, the objectives in the spatial planning of land use are proposed at two levels: at the municipal overall land use plan level, and at the urban master plan level. The objectives of these two levels consider different problems that are based on their own functions. In the phase of assigning the objectives for Shenzhen, three land use principles, national and local policies, and the characteristics of Shenzhen are considered. The objectives can be classified as spatial land use allocation-related and land use structure-related. For land use structure-related objectives, a coefficient approach is usually used to measure the objectives. For example, the economic benefit and the service value are calculated by timing the area of a certain land use type and the GDP coefficient and ecosystem service coefficient, respectively. The difficulty in measuring these objectives depends on the identification of the specific coefficients. However, the process of measuring spatially related objectives is difficult because of the complex, nonlinear, and unstructured features of spatial problems.

Using the MOO, a set of Pareto solutions from which the planner can select is derived. Firstly, the MOO generates the Pareto solutions at the municipal overall land use level, which determines the built-up extent in the projected year. Then, the MOO conducts the spatial planning of land use at the urban master plan level based on the determined built-up extent in the first step. Using this two-level MOO approach, coordination between the municipal overall land use plan and the urban master plan is achieved. However, planners and scholars believe that forecasting the most important key index in the plans, such as the population and total area of built-up land in the projected year, is difficult. The MOO proposes a potential solution for this difficulty by providing a set of Pareto solutions as alternative plans. If the set constraint on the area of built-up land is an interval rather than a specific value, then the alternative plan maintains a series of areas of built-up land within the given interval. Within this context, the planners will not be apprehensive of the rigorous constraints on the built-up land.

Over all, this book has made some significant contributions to the MOO study field.

A computer was used to measure the relatively complex land use objectives. A computer-based measurement improves our understanding of the effects of spatial land use change on economic development and environmental preservation. With this knowledge, arranging and managing land use resources becomes efficient.

The spatial planning of land use focuses on the spatial layout of land use and is a core part of land use planning. In this study, the spatial planning of land use for Shenzhen was conducted at two levels to facilitate coordination between the municipal overall land use plan and the urban master plan. Difficulties in the implementation of spatial land use layout to achieve the proposed goals or objectives of land use planning are confirmed and discussed. This study focuses on the spatial inconsistencies and aims to distribute urban land uses based on the spatial distribution of municipal overall land uses. Even if the inconsistencies between the municipal overall land use plan and the urban master plan originate from many aspects, this study considers such spatial inconsistencies and drafts a solution.

This approach helps in manual planning. In the process of manual planning, planners need to first set up the objectives, and then determine whether all these objectives are considered in the plan. Firstly, evaluating the objectives helps planners to understand the effects of land use changes on the objectives. Through the evaluation function, planners can easily predict what will happen if the land use changes in a specific manner. Secondly, the model ranks the alternative plans by fitness function, which helps the planners to compare the positive and negative aspects of potential plans. Finally, the approach provides a set of alternative plans from which planners and policy-makers can select. These plans can be used as a draft or a solution and can be modified by the planners. The Pareto solutions can be considered to be the land use scenarios, which can benefit the planners if they know how to accomplish the land use objectives under each scenario as provided by the MOO used in this study.

Meanwhile, this study considered multiple land use objectives, which reflect the goals or targets of land use planning. The objectives are proposed at two levels: at the municipal overall land use plan level, and at the urban master plan level. This idea is innovative and practical in the Chinese context because numerous plans at different levels comprise the land use planning system in China. Moreover, a number of complex spatially related objectives, which are considered in the MOO approach, are proposed. These spatial objectives can be measured by using GIS and RS.

References

Abu Hammad, A. 2011. Watershed Erosion Risk Assessment and Management Utilizing Revised Universal Soil Loss Equation-geographic Information Systems in the Mediterranean Environments. Water and Environment Journal, Vol. 25, No. 2, pp. 149–162.

Aerts, J.C., Eisinger, E., Heuvelink, G., & Stewart, T.J. 2003. Using Linear Integer Programming for Multi-site Land-Use Allocation. Geographical Analysis, Vol. 35, No. 2, pp. 148–169.

Agarwal, C., Green, G.M., Grove, J.M., Evans, T.P., & Schweik, C.M. 2002. A Review and Assessment of Land-use Change Models: Dynamics of Space, Time, and Human Choice. Gen. Tech. Rep. NE-297. Newton Square, PA, U.S. Department of Agriculture, Forest Service, Northeastern Research Station. 61.

Ahangaran, M., Taghizadeh, N., & Beigy, H. 2017. Associative Cellular Learning Automata and its Applications. Applied Soft Computing, Vol. 53, pp. 1–18.

Akiva, M.E.B. & Lerman, S.R. 1985. Discrete Choice Analysis: Theory and Application to Predict Travel Demand. The MIT Press.

Alessandro, P. & Gerald, C.N. 2009. Land Use Change with Spatially Explicit Data: A Dynamic Approach. Environmental and Resource Economics, Vol. 43, No. 2, pp. 209–229.

Alexander, P. Katherine, C., et al. 2016. Land use futures in the shared socio-economic pathways. Global Environmental Change, Vol. 10, pp. 112–121.

Alig, R.J., Kline, J.D., & Lichtenstein, M. 2004. Urbanization on the US Landscape: Looking Ahead in the 21st Century. Landscape and Urban Planning, Vol. 69, pp. 214–234.

Alonso, W. 1968. Equity and its Relation to Efficiency in Urbanization. Institute of Urban & Regional Development, University of California.

Alonso, D. & Sole, R.V. 2000. The DivGame Simulator: A stochastic Cellular Automata Model of Rainforest Dynamics. Ecological Modelling, Vol. 133, No. 1/2, pp. 131–141.

An, L. & Brown, D.G. 2008. Survival Analysis in Land-change Science: Integrating with GIScience to Address Temporal Complexities. Annals of Association of American Geographers, Vol. 98, No. 2, pp. 1–22.

An, Q. 2013. Boosting Smart Growth Planning in Shenzhen by Spatial Planning. Practice and Theory of Seas, Vol. 6, pp. 80–83. (In Chinese).

Andreoni, J.A. & Miller, J.H. 1993. Rational Cooperation in the Finitely Repeated Prisoner's Dilemma: Experimental Evidence. Economic Journal, Vol. 103, No. 418, pp. 570–585.

Anselin, L. 1988. Spatial Econometrics: Methods and Models. Dordrecht, Kluwer Academic. 284.

Anselin, L. & Bera, A. 1998. Spatial Dependence in Linear Regression Models with an Introduction to Spatial Econometrics. In A. Ullah & D. Giles (eds.) Handbook of Applied Economic Statistics. New York, Marcel Dekker, 237–289.

Anselin, L. 1998. GIS Research Infrastructure for Spatial Analysis of Real Estate Markets. Journal of Housing Research, Vol. 9 pp. 113–133.

Arad, R.W. & Berechman, J. 1978. A Design Model for Allocating Interrelated Land-use Activities in Discrete Space. Environment and Planning A. Vol. 10, No. 11, pp. 1319–1332.

AUMA, 2007. Summit Land Use Planning Background Paper.

Auyang, S.Y. 1998. Foundations of Complex Systems Theories: In Economics, Evolutionary Biology, and Statistical Physics. Cambridge University Press, Cambridge, New York, Melbourne.

Axtell, R.L. & Epstein, J.M. 1994. Agent-based Modeling: Understanding our Creations. The Bulletin of the Santa Fe Institute, pp. 28–32.

Baker, W.L. 1989. A Review of Models in Landscape Change. Landscape Ecology, Vol. 2, No. 2, pp. 111–133.

Baker, M., Coaffee, J., & Sherriff, G. 2007. Achieving Successful Participation in the New UK Spatial Planning System. Planning, Practice & Research, Vol. 22, No. 1, pp. 79–93.

Bakker, N., Dubbeling, M., Gündel, S., Sabel Koschella, U., & Zeeuw, H.D. 2000. Growing Cities, Growing Food: Urban Agriculture on the Policy Agenda. A reader on urban agriculture, DSE.

Balling, R.J., Taber, J.T., Brown, M., & Day, K. 1999. Multiobjective Urban Planning Using a Genetic Algorithm. ASCE Journal of Urban Planning and Development, Vol. 125, No. 2, pp. 86–99. 53.

Balling, R. 2002. The Maximin Fitness Function for Multiobjective Evolutionary Optimization. Springer, Optimization in Industry. 135–147.

Balling, R. 2003. The Maximin Fitness Function; Multi-objective City and Regional Planning. Springer, Evolutionary Multi-Criterion Optimization.

Balling, R., Powell, B., & Saito, M. 2004. Generating Future Land–Use and Transportation Plans for High–Growth Cities Using a Genetic Algorithm. Computer–Aided Civil and Infrastructure Engineering, Vol. 19, No. 3, pp. 213–222.

Balzter, H., Braun, P.W., & Kohler, W. 1998. Cellular Automata Models for Vegetation Dynamics. Ecological Modelling, Vol. 107, No. 2/3, pp. 113–125.

Bammi, D. & Bammi, D. 1979. Development of a Comprehensive Land use Plan by Means of a Multiple Objective Mathematical Programming Model. Interfaces, Vol. 9, No. 2-Part-2, pp. 50–63.

Barber, G. 1976. Land-use Plan Design via Interactive Multiple-objective Programming. Environment and Planning A, Vol. 8, No. 6, pp. 625–636.

Başkent, E.Z. & Kadıoğulları, A.İ. 2007. Spatial and Temporal Dynamics of Land use Pattern in Turkey: A Case Study in İnegöl. Landscape and Urban Planning, Vol. 81, No. 4, pp. 316–327.

Batty, M., Xie, Y., & Sun, Z. 1999. Modeling Urban Dynamics through GIS-based Cellular Automata. Computers, Environment and Urban Systems, Vol. 23, No. 3, pp. 205–233.

Batty, M. 2001. Agent-based Pedestrian Modeling. Environment and Planning B, Vol. 28, No. 3, pp. 321–326.

Benayoun, R., De Montgolfier, J., Tergny, J., & Laritchev, O. 1971. Linear Programming with Multiple Objective Functions: Step Method (STEM). Mathematical Programming, Vol. 1, No. 1, pp. 366–375.

Bendor, T.K., Spurlock, D., Woodruff, S.C., & Olander, L. 2016. A Research Agenda for Ecosystem Services in American Environmental and Land use Planning. Cities, Vol. 60, pp. 260–271.

Benenson, I. 1998. Multiagent Simulations of Residential Dynamics in the City. Computers, Environment and Urban Systems, Vol. 22, No. 1, pp. 25–42.

Berger, T. 2001. Agent-based Spatial Models Applied to Agriculture: A Simulation Tool for Technology Diffusion, Resource use Changes, and Policy Analysis. Agricultural Economics, Vol. 25, No. 2–3, pp. 245–260.

Bielecki, T.R., Jakubowski, J., & Niewęgłowski, M. 2016. Conditional Markov chains: Properties, Construction and Structured Dependence. Stochastic Processes & Their Applications.

Bierens, H.J. & Hoever, R. 1985. Population Forecasting at the City Level: An Econometric Approach. Urban Studies, Vol. 22, No. 1, pp. 83–90.

Biłozor, A., Renigierbiłozor, M., & Furman, A. 2014. Multi-Criteria Land use Function Optimization. Real Estate Management & Valuation, Vol. 22, No. 4, pp. 81–91.

Blaschke, P.M., Dickinson, K.J.M., & Roper-Lindsay, J. 1991. Defining sustainability – is it worth it? Proceedings of the International Conference on Sustainable Land Management. Napier, New Zealand.

Bockstael, N. 1996. Modeling Economics and Ecology: The Importance of a Social Perspective. American Journal of Agricultural Economics, Vol. 78, No. 5, pp. 1168–1180.

Brookes, C.J. 2001. A Genetic Algorithm for Designing Optimal Patch Configurations in GIS. International Journal of Geographical Information Science, Vol. 15, No. 6, pp. 539–559.

Brown, D.G., Pijanowski, B.C., & Duh, J.D. 2000. Modeling the Relationships between Land use and Land Cover on Private Lands in the Upper Midwest, USA. Journal of Environmental Management, Vol. 59, pp. 247–263.

Brown, D.G., Goovaerts, P., Burnicki, A., & Li, M.Y. 2002. Stochastic Simulation of Land-cover Change Using Geostatistics and Generalized Additive Models. Photogrammetric Engineering & Remote Sensing, Vol. 68, No. 10, pp. 1051–1061.

Brunsdon, C., Fotheringham, A.S., & Charlton, M. 1996. Geographically Weighted Regression: A Method for Exploring Spatial Non-stationarity, Geographical Analysis, Vol. 28, pp. 281–298.

Brunsdon, C., Fotheringham, A.S., & Charlton, M. 1999. Some Notes on Parametric Significance Tests for Geographically Weighted Regression. Journal of Regional Sciences, Vol. 39, No. 3, pp. 497–524.

Bruton, M.J., Bruton, S.G., & Li, Y. 2005. Shenzhen: Coping with Uncertainties in Planning. Habitat International, Vol. 29, No. 2 pp. 227–243.

Cai, C., Ding, S., Shi, Z., Huang, L., & Zhang, G. 2000. Study of Applying USLE and Geographical Information System IDRISI to Predict Soil Erosion in Small Watershed. Journal of Soil and Water Conservation, Vol. 14, No. 2, pp. 19–24.

Can, A., 1992. Specification and Estimation of Hedonic Housing Price Models. Regional Science and Urban Economics, Vol. 22 pp. 453–474.

Can, A. & Megbolugbe, I. 1997. Spatial Dependence and House Price Index Construction. Journal of Real Estate Finance and Economics, Vol. 14, pp. 203–222.

Cao, K., Batty, M., Huang, B., Liu, Y., Yu, L., & Chen, J. 2011. Spatial Multi-objective Land use Optimization: Extensions to the Non-dominated Sorting Genetic Algorithm-II. International Journal of Geographical Information Science, Vol. 25, No. 12, pp. 1949–1969.

Chandramouli, M. 2007. Integration of GA-based Multiobjective Optimization with VR-based Visualization to Solve Landuse Problems. Masters Abstracts International.

Chandramouli, M., Huang, B., & Xue, L. 2009. Spatial Change Optimization: Integrating GA with Visualization for 3D Scenario Generation. Photogrammetric Engineering and Remote Sensing, Vol. 75, No. 8, pp. 1015–1023.

Chang, Q., Li, X., Huang, X., & Wu, J. 2012. A GIS-based Green Infrastructure Planning for Sustainable Urban Land Use and Spatial Development. 2011 International Conference of Environmental Science and Engineering, Vol. 12, Pt A. M. Ma. 12, pp. 491–498.

Chazan, D., Cotter, A.A., Keoleian, G.A., & Marans, R.W. 2001. Evaluating the Impacts of Proposed Land Conversion: A Tool for Local Decision-Making, Citeseer.

Chen, S.H. & Yeh, C.H. 2001. Evolving Traders and the Business School with Genetic Programming: A New Architecture of the Agent-based Artificial Stock Market. Journal of Economic Dynamics and Control, Vol. 25, No. 3–4, pp. 363–393.

Chen, Y., Zhang, Z., Du, S., Shi, P., Tao, F., & Doyle, M. 2011. Water Quality Changes in the World's First Special Economic Zone, Shenzhen, China. Water Resources Research, Vol. 47, No. 11.

Chen, Z., Gong, C., Wu, J., & Yu, S. 2012. The Influence of Socioeconomic and Topographic Factors on Nocturnal Urban Heat Islands: A Case Study in Shenzhen, China. International Journal of Remote Sensing, Vol. 33, 12 pp. 3834–3849.

Cheng, J. & Masser, I., 2003, Urban Growth Modeling: A Case Study of Wuhan city, PR China. Landscape and Urban Planning, Vol. 62, pp. 199–217.

Chuai, X., Huang, X.J., Wu, C., et al. 2016. Land use and Ecosystems Services Value Changes and Ecological Land Management in Coastal Jiangsu, China. Habitat International.

Clarke, K.C. & Gaydos, L.J. 1998, Loose-coupling a CA model and GIS: Long-term Urban Growth Prediction for San Franciso and Washington/Baltimore. International Journal of Geographical Information Science, Vol. 12, pp. 699–714.

Costanza, R., d'Arge, R., de Groot, R., Farber, S., Grasso, M., Hannon, B., Limburg, K., Naeem, S., O'Neill, R.V., & Paruelo, J. 1998. The Value of the World's Ecosystem Services and Natural Capital.

Dai, F., Lee, C., & Zhang, X. 2001. GIS-based Geo-Environmental Evaluation for Urban Land-use Planning: A Case Study. Engineering Geology, Vol. 61, No. 4, pp. 257–271.

Dalvi, M.Q. & Martin, K. 1976. The Measurement of Accessibility: Some Preliminary Results. Transportation, Vol. 5, No. 1, pp. 17–42.

Daniels, T. 2001. Smart Growth: A New American Approach to Regional Planning. Planning Practice and Research, Vol. 16, No. 3–4, pp. 271–279.

Datta, R. & Regis, R.G. 2016. A Surrogate-Assisted Evolution Strategy for Constrained Multi-Objective Optimization. Expert Systems with Applications, Vol. 57, pp. 270–284.

DeRose, R. & North, P. 1995. Slope Limitations to Sustainable Land use in Hill Country Prone to Landslide Erosion. Unpublished Report, Landcare Research, Palmerston North, New Zealand.

Dhakal, S. 2009. Urban Energy use and Carbon Emissions from Cities in China and Policy Implications. Energy Policy, Vol. 37, No. 11 pp. 4208–4219.

Dillon, P. & Kirchner, W. 1975. The Effects of Geology and Land use on the Export of Phosphorus from Watersheds. Water Research, Vol. 9, No. 2 pp. 135–148.

Dombrow, J., Knight, J.R., & Sirmans, C.F. 1997. Aggregation Bias in Repeat Sales Indices. Journal of Real Estate Finance and Economics, Vol. 14, pp. 75–88.

Douvere, F. 2008. The Importance of Marine Spatial Planning in Advancing Ecosystem-based Sea use Management. Marine Policy, Vol 32, No. 5, pp. 762–771.

Downs, A. 2005. Smart Growth: Why we iscuss it more than we do it. Journal of the American Planning Association, Vol. 71, No. 4, pp. 367–378.

Dubin, R.A. 1992. Spatial Correlation and Neighborhood Quality. Regional Science and Urban Economics, Vol. 22, pp. 432–452.

Dubin, R. 1995. Estimating Logit Models with Spatial Dependence. in New Directions in Spatial Econometrics. In L. Anselin & R. Florax (eds.), New York, Springer-Verlag.

Dumas, P., Printemps, J., Mangeas, M., & Luneau, G. 2010. Developing Erosion Models for Integrated Coastal Zone Management: A Case Study of the New Caledonia West Coast. Marine Pollution Bulletin, Vol. 61, No. 7 pp. 519–529.

Eckert, J.K. (ed.). 1990. Property Appraisal and Assessment Administration. Chicago, Illinois, International Association of Assessing Officers.

Fang, J., Guo, Z., Piao, S., & Chen, A. 2007. Terrestrial Vegetation Carbon Sinks in China, 1981–2000. Science in China Series D: Earth Sciences, Vol. 50, No. 9 pp. 1341–1350.

FAO, 1993. Guidelines for Land-Use Planning, Food and Agriculture Organization of The United Nations Rome.

Feng, C.M. & Lin, J.J. 1999. Using a Genetic Algorithm to Generate Alternative Sketch Maps for Urban Planning. Computers, Environment and Urban Systems, Vol. 23, No. 2, pp. 91–108.

Foody, G.M. 2004. Thematic Map Comparison: Evaluating the Statistical Significance of differences in Classification Accuracy. Photogrammetric Engineering & Remote Sensing, Vol. 70, No. 5, pp. 627–633.

Fotheringham, A.S., Charlton M.E., & Brunsdon, C. 1996. The Geography of Parameter Space: An Investigation of Spatial Non-stationarity. International Journal of Geographical Information Science, Vol. 10, pp. 605–627.

Fotheringham, A.S., Charlton, M.E., & Brunsdon, C. 1998. Geographically Weighted Regression: A Natural Evolution of the Expansion Method for Spatial Data Analysis. Environment and Planning A, Vol. 30, pp. 1905–1927.

Fotheringham, A.S., Brunsdon C., & Charlton, M.E. 2002. Geographically Weighted Regression: The Analysis of Spatially Varying Relationships. Chichester: Wiley.

Gabriel, S.A., Faria, J.A., & Moglen, G.E. 2006. A Multiobjective Optimization Approach to Smart Growth in Land Development. Socio-Economic Planning Sciences, Vol. 40, No. 3, pp. 212–248.

Gelfand, A.E., Ecker, M.D., Knight, J.R., & Sirmans, C.F. 2001. The Dynamics of Location in Home Prices. Working Paper, University of Connecticut.

Gelfand, A.E., Kim, H-J., Sirmans, C.F., & Banerjee, S. 2003. Spatial Modeling with Spatially Varying Coefficients Processes. Journal of the American Statistical Association, Vol. 98, pp. 387–396.

Geoghegan, J., Villar, S.C., Klepeis, P., et al. 2001. Modeling Tropical Deforestation in the Southern Yucatan Peninsular Region: Comparing Survey and Satellite Data. Agriculture, Ecosystems and Environment, Vol. 85, pp. 25–46.

Gill, J. & King, G. 2004. What to do When Your Hessian is not Invertible: Alternatives to Model Respecification in Nonlinear Estimation. Sociological Methods and Research, Vol. 32, No. 1, pp. 54–87.

Gilruth, P.T., Marsh, S.E., & Itami, R. 1995. A Dynamic Spatial Model of Shifting Cultivation in the Highlands of Guinea, West Africa. Ecological Modelling, Vol. 79 pp. 179–197.

Gimblett, H.R., (ed.) 2002. Integrating Geographic Information Systems and Agent-Based Modeling Techniques for Simulating Social and Ecological Processes. New York, Oxford University Press.

Glade, T. 2003. Landslide Occurrence as a Response to Land use Change: A Review of Evidence from New Zealand. Catena, Vol. 51, No. 3, pp. 297–314.

Goodman, A.C. & Thibodeau, T.G. 1998, Housing Market Segmentation. Journal of Housing Economics, Vol. 7, pp. 121–143.

Greenberg, M.R. 1978. Applied Linear Programming for the Socioeconomic and Environmental Sciences. Academic Press.

Greene, W. 2000. Econometric Analysis. Upper Saddle River: Prentice-Hall.

Grimmond, C. & Souch, C. 1994. Surface Description for Urban Climate Studies: A GIS Based Methodology. Geocarto International, Vol. 9, No. 1, pp. 47–59.

Guido, S., Federica, M., Chiara, M., & Stefano, C. 2016. Exploring Land Use Scenarios in Metropolitan Areas: Food Balance in a Local Agricultural System by Using a Multi-objective Optimization Model. Agriculture & Agricultural Science Procedia, Vol. 8, pp. 211–221.

Gumpertz, M.L., Wu, C., & Pye, J.M. 2000. Logistic Regression for Southern Pine Beetle Outbreaks with Spatial and Temporal Correlation. Forest Science, Vol. 46, No. 1, pp. 95–107.

Hajela, P. & Lin, C.-Y. 1992. Genetic search strategies in multicriterion optimal design. Structural Optimization, Vol. 4, No. 2, pp. 99–107.

Hansen, W.G. 1959. How accessibility shapes land use. Journal of the American Institute of Planners, Vol. 25, No. 2, pp. 73–76.

Harbor, J. 1999. Engineering Geomorphology at the Cutting Edge of Land Disturbance: Erosion and Sediment Control on Construction Sites. Geomorphology, Vol. 31, No. 1, pp. 247–263.

He, F.L., Zhou, J., & Zhu, H. 2003. Autologistic Regression Model for the Distribution of Vegetation. Journal of Agricultural, Biological, and Environmental Statistics, Vol. 8, No. 2, pp. 205–222.

Hopkins, L. 1977. Land-use Plan Design—Quadratic Assignment and Central-facility Models. Environment and Planning A, Vol. 9, No. 6, pp. 625–642.

Hosmer, D. & Lemeshow, S. 1989. Applied Logistic Regression. New York, John Wiley and Sons, Inc.

Huang, B. & Claramunt, C. 2005. Spatiotemporal Data Model and Query Language for Tracking Land use Change. Transportation Research Record: Journal of the Transportation Research Board, Vol. 1902, pp. 107–113.

Huang, B., Fery, P., Xue, L., & Wang, Y. 2008. Seeking the Pareto Front for Multiobjective Spatial Optimization Problems. International Journal of Geographical Information Science, Vol. 22, No. 5, pp. 507–526.

Huang, B., Zhang, L., & Wu, B. 2009a. Spatio-temporal Analysis of Rural-urban Land Conversion. International Journal of Geographical Information Science, Vol. 23, No. 3 pp. 379–398.

Huang, B., Xie, C.L., & Tay, R. 2009b. Land use Change Modeling using Unbalanced Support Vector Machines. Environment and Planning B, Vol. 36, No. 3, pp. 398–416.

Huang, B., Wu, B., & Barry, M. 2010. Geographically and Temporally Weighted Regression for Modeling Spatio-temporal Variation in House Prices. International Journal of Geographical Information Science, Vol. 24, No. 3, pp. 383–401.

Huffer, F.W. & Wu, H.L. 1998. Markov chain Monte Carlo for Auto-logistic Regression Models with Application to the Distribution of Plant Species. Biometrics, Vol. 54, pp. 509–524.

Hurvich, C.M., Simonoff, J.S., & Tsai C.L. 1998. Smoothing Parameter Selection in Nonparametric Regression Using an Improved Akaike Information Criterion. Journal of the Royal Statistical Society (Series B), Vol. 60, pp. 271–293.

Irwin, E.J. & Geohegan J. Theory, data, methods: Developing spatially explicit economic models of land use change. Agriculture, Ecosystems, and Environment, Vol. 85, pp. 7–24.

Jiang, Y., Fu, M.C., Wang, Z., Zhang, Z.Y., Song, B.H., & Wen, H.Y. 2010. Impacts of Land use Change on Ecosystem Carbon Sinks and Sources: Wuan, Hebei as a Case Study. Journal of Anhui Agricultural Science, Vol. 38, No. 24, pp. 13067–13069.

Johnes, P.J. 1996. Evaluation and Management of the Impact of Land use Change on the Nitrogen and Phosphorus Load Delivered to Surface Waters: The Export Coefficient Modelling Approach. Journal of Hydrology, Vol. 183, No. 3 pp. 323–349.

Journal, A. & Huijbregis C.H. 1978. Mining geostatisticals. London, UK, Academic Press.

Kafadar, K. 1994. Choosing Among Two-dimensional Smoothers in Practice. Comp. Stat. & Data Analysis, Vol. 18, pp. 419–439

Kaimowitz, D. & Angelsen, A. 1998. Economic Models of Tropical Deforestation: A Review. Centre for International Forestry Research, Jakarta, Indonesia.

Kaur, R., Srivastava, R., Betne, R., Mishra, K., & Dutta, D. 2004. Integration of Linear Programming and a Watershed-scale Hydrologic Model for Proposing an Optimized Land-use Plan and Assessing its Impact on Soil Conservation—A Case Study of the Nagwan Watershed in the Hazaribagh District of Jharkhand, India. International Journal of Geographical Information Science, Vol. 18, No. 1, pp. 73–98.

Kay, D., Wyer, M.D., Crowther, J., Wilkinson, J., Stapleton, C., & Glass, P. 2005. Sustainable Reduction in the Flux of Microbial Compliance Parameters from Urban and Arable Land use to Coastal Bathing Waters by a Wetland Ecosystem Produced by a Marine Flood Defence Structure. Water Research, Vol. 39, No. 14, pp. 3320–3332.

Khadam, I.M. & Kaluarachchi, J.J. 2006. Water Quality Modeling under Hydrologic Variability and Parameter Uncertainty using Erosion-scaled Export Coefficients. Journal of Hydrology, Vol. 330, No. 1, pp. 354–367.

Kollman, K., Miller, J.H., & Page, S.E. 1997. Political Institutions and Sorting in a Tiebout Model. American Economic Review, Vol. 87, No. 5, pp. 977–992.

Konak, A., Coit, D.W., & Smith, A.E. 2006. Multi-objective Optimization using Genetic Algorithms: A Tutorial. Reliability Engineering & System Safety, Vol. 91, No. 9, pp. 992–1007.

Koster, H.R. & Rouwendal, J. 2012. The Impact of Mixed Land use on Residential Property Values*. Journal of Regional Science, Vol. 52, No. 5, pp. 733–761.

Krider, R.E. & Weinberg, C.B. 1997. Spatial Competition and Bounded Rationality: Retailing at the Edge of Chaos. Geographical Analysis, Vol. 29, No. 1, pp. 16–34.

Kumar, S. & Kushwaha, S. 2013. Modeling Soil Erosion Risk based on RUSLE-3D using GIS in a Shivalik sub-watershed. Journal of Earth System Science, pp. 1–10.

Lagarias, J.C., Reeds, J.A., Wright, M.H., & Wright, P.E. 1998. Convergence Properties of the Nelder-Mead Simplex Method in Low Dimensions, SIAM Journal of Optimization, Vol. 9, No. 1, pp. 112–147, 1998.

Lambin, E.F., Geist, H.J., & Lepers, E. 2003. Dynamics of Land-use and Landcover Change in Tropical Regions. Annual Review Environment Resource, Vol. 28, pp. 205–241.

Landry, S.M. & Chakraborty, J. 2009. Street Trees and Equity: Evaluating the Spatial Distribution of an Urban Amenity. Environment and Planning A, Vol. 41, No. 11, pp. 2651.

Langville, A.N. & Steward W.J. 2004. The Kronecker Product and Stochastic Automata Networks. Journal of Computational and Applied Mathematics, Vol. 164, pp. 429–447.

Lee, S. 2004. Soil Erosion Assessment and its Verification using the Universal Soil Loss Equation and Geographic Information System: A case study at Boun, Korea. Environmental Geology, Vol. 45, No. 4, pp. 457–465.

Lee, S. & Talib, J.A. 2005. Probabilistic Landslide Susceptibility and Factor Effect Analysis. Environmental Geology, Vol. 47, No. 7, pp. 982–990.

Lemly, A.D. 1982. Erosion control at construction sites on red clay soils. Environmental Management, Vol. 6, No. 4, pp. 343–352.

Leung, Y., Mei, C.L., & Zhang, W.X. 2000. Statistical Tests for Spatial Non-stationarity Based on the Geographically Weighted Regression Model. Environment and Planning A, Vol. 32, pp. 9–32.

Leung, H.-L. 2003. Land use Planning Made Plain. University of Toronto Press.

Li, X. & Parrott, L. 2016. An improved Genetic Algorithm for Spatial Optimization of Multi-objective and Multi-site Land use Allocation. Computers Environment & Urban Systems, Vol. 59, pp. 184–194.

Li, L., Chen, C., Xie, S., Huang, C., Cheng, Z., Wang, H., Wang, Y., Huang, H., Lu, J., & Dhakal, S. 2010. Energy Demand and Carbon Emissions under Different Development Scenarios for Shanghai, China. Energy Policy, Vol. 38, No. 9, pp. 4797–4807.

Li, T., Li, W., & Qian, Z. 2010. Variations in Ecosystem Service Value in Response to Land use Changes in Shenzhen. Ecological Economics, Vol. 69, No. 7, pp. 1427–1435.

Ligmann-Zielinska, A., Church, R.L., & Jankowski, P. 2008. Spatial Optimization as a Generative Technique for Sustainable Multiobjective Land-use Allocation. International Journal of Geographical Information Science, Vol. 22, No. 6, pp. 601–622.

Lindsey, G., Maraj, M., & Kuan, S. 2001. Access, Equity, and Urban Greenways: An Exploratory Investigation. The Professional Geographer, Vol. 53, No. 3, pp. 332–346.

Liu, R.M., Yang, Z.F., Shen, Z.Y., Yu, S.L., Ding, X.W., Wu, X., & Liu, F. 2009. Estimating Nonpoint Source Pollution in the Upper Yangtze River using the Export Coefficient Model, Remote Sensing, and Geographical Information System. Journal of Hydraulic Engineering, Vol. 135, No. 9, pp. 698–704.

Liu, W.C. 1995. Soil and Water Loss in the Process of Construction and Development of Shenzhen and its Treatment. Journal of Nanchang College of Water Conservancy and Hydroelectric Power, Vol. s1, pp. 74–77.

Liu, Y., Liu, D., Liu, Y., He, J., Jiao, L., Chen, Y., & Hong, X. 2012. Rural Land use Spatial Allocation in the Semiarid Loess Hilly Area in China: Using a Particle Swarm Optimization Model Equipped with Multi-objective Optimization Techniques. Science China-Earth Sciences, Vol. 55, No. 7, pp. 1166–1177.

Liu, Y., Wang, H., Ji, Y., Liu, Z., & Zhao, X. 2012. Land Use Zoning at the County Level Based on a Multi-Objective Particle Swarm Optimization Algorithm: A Case Study from Yicheng, China. International Journal of Environmental Research and Public Health, Vol. 9, No. 8, pp. 2801–2826.

Longley, P., Higgs, G., & Martin, D. 1994. The Predictive use of GIS to Model Property Valuations. International Journal of Geographical Information Systems, Vol. 8, No. 2, pp. 217–235.

Lowry, M.B. & Balling, R.J. 2009. An Approach to Land-use and Transportation Planning that Facilitates City and Region Cooperation. Environment and Planning B: Planning and Design, Vol. 36, No. 3, pp. 487–504.

Lu, W.J., Yang, L.M., & Li, Y.X. 2004. An Analysis on the Coordination Between City Planning and Comprehensive Land use Planning. City Planning Review, Vol. 4, pp. 58–63. (In Chinese).

Ludeke, A.K., Maggio, R.C., & Reid, L.M. 1990. An Analysis of Anthropogenic Deforestation using Logistic Regression and GIS. Journal of Environmental Management, Vol. 31, pp. 247–259. 66.

Lufafa, A., Tenywa, M., Isabirye, M., Majaliwa, M., & Woomer, P. 2003. Prediction of Soil Erosion in a Lake Victoria Basin Catchment using a GIS-based Universal Soil Loss model. Agricultural Systems, Vol. 76, No. 3, pp. 883–894.

Luo, D. & Zhang, W. 2013. A Comparison of Markov Model-based Methods for Predicting the Ecosystem Service Value of Land use in Wuhan, Central China. Ecosystem Services.

Lv, Z.Q., Wu, Z.F., Wei, J.B., Sun, C., Zhou, Q.G., & Zhang, J.H. 2011. Monitoring of the Urban Sprawl using Geoprocessing Tools in the Shenzhen Municipality, China. Environmental Earth Sciences, Vol. 62, No. 6, pp. 1131–1141.

Ma, C., Ma, J., & Buhe, A. 2001. Quantitative Assessment of Vegetation Coverage Factor in USLE Model Using Remote Sensing Data. Bulletin of Soil and Water Conservation, Vol. 21, No. 4, pp. 6–9.

Ma, H., Jiang, J., Song J., et al. 2016. Effects of Urban Land-use Change in East China on the East Asian Summer Monsoon Based on the CAM5.1 model. Climate Dynamics, Vol. 46, No. 9–10, pp. 2977–2989.

Manard, S. 1995. Applied Logistic Regression Analysis. Series: Quantitative Applications in the Social Sciences. Thousand Oaks, CA, Sage.

Manel S., Dias, J.M., & Ormerod, S.J. 1999. Comparing Discriminant Analysis, Neural Networks and Logistic Regression for Predicting Species' Distribution: A Case Study with a Himalayan River Bird. Ecological Modelling, Vol. 120, pp. 337–347.

Manolopoulos, H., Snyder, D.C., Schauer, J.J., Hill, J.S., Turner, J.R., Olson, M.L., & Krabbenhoft, D.P. 2007. Sources of Speciated Atmospheric Mercury at a Residential Neighborhood Impacted by Industrial Sources. Environmental Science & Technology, Vol. 41, No. 16, pp. 5626–5633.

Marimon, R., McGratten, E., & Sargent, T.J. 1990. Money as a Medium of Exchange in an Economy with Artificially Intelligent Agents. Journal of Economic Dynamics and Control, Vol. 14, pp. 329–373.

Marks, R.E. 1992. Breeding Hybrid Strategies: Optimal Behavior for Oligopolists. Journal of Evolutionary Economics, Vol. 2, pp. 17–38.

Maruyama, T. & Sumalee, A. 2007. Efficiency and Equity Comparison of Cordon-and Area-based Road Pricing Schemes Using a Trip-chain Equilibrium Model. Transportation Research Part A: Policy and Practice, Vol. 41, No. 7, pp. 655–671.

Masoomi, Z., Mesgari, M.S., & Hamrah, M. 2013. Allocation of urban land uses by Multi-Objective Particle Swarm Optimization algorithm. International Journal of Geographical Information Science, Vol. 27, No. 3, pp. 542–566.

McCool, D., Brown, L., Foster, G., Mutchler, C., & Meyer, L. 1987. Revised Slope Steepness Factor for the Universal Soil Loss Equation. American Society of Agricultural Engineers, Transactions TAAEAJ, Vol. 30, No. 5.

McMillen, D.P. 1996. One Hundred Fifty Years of Land Values in Chicago: A Nonparametric Approach. Journal of Urban Economics, Vol. 40, pp. 100–124.

McMillen, D.P. & McDonald J.F. 1997. A Nonparametric Analysis of Employment Density in a Polycentric City. Journal of Regional Science, Vol. 37, pp. 591–612.

Meusburger, K., Konz, N., Schaub, M., & Alewell, C. 2010. Soil Erosion Modelled with USLE and PESERA Using QuickBird Derived Vegetation Parameters in an Alpine Catchment. International Journal of Applied Earth Observation and Geoinformation, Vol. 12, No. 3, pp. 208–215.

Miettinen, K. 1999. Nonlinear Multiobjective Optimization. Springer.

Mika, A.M., Jenkins, J.C., Hathaway, K., Lawe, S., & Hershey, D. 2010. Vermont Integrated Land use and Transportation Carbon Estimator. Transportation Research Record: Journal of the Transportation Research Board, Vol. 2191, No. 1, pp. 119–127.

Miller, J.H. & Shubik, M. 1992. Some Dynamics of a Strategic Market Game with a Large Number of Agents. Santa Fe, NM: Santa Fe Institute Publication 92–11-057.

Morotti, S. & Grandi, E. 2016. Logistic Regression Analysis of Populations of Electrophysiological Models to Assess Proarrythmic Risk. Methodsx, Vol. 4 pp. 25–34.

Mu, T.L., Xie, J., Wu, J.S., Wang, X.R., & Zheng, M.K. 2010. Effects of Land Use on Soil Erosion in Shenzhen, China. Research of Soil and Water Conservation, Vol. 3, pp. 012.

Munroe, D.K., Southworth, J., & Tucker, C.M. 2004. Modelling Spatially and Temporally Complex Land-Cover Change: The Case of Western Honduras. The Professional Geographer, Vol. 56, No. 4, pp. 544–559.

Nachtergaele, F., van Velthuizen H., & Verelst, L. 2008. Harmonized World Soil Database. F. a. A. O. o. t. U. Nations.

Natural Resources Conservation Service, United States Department of Agriculture. 1999. Summary Report 1997 Natural Resources Inventory. Web http://www.nhq.nrcs.usda.gov/NRI, accessed on March 20, 09.

Neci, J.T., Kirby, K.R., Gilbert, B., & Starzomski, B.M. 2016. The Impact of Land-use Change on Larval Insect Communities: Testing the Role of Habitat Elements in Conservation. Ecoscience, Vol. 15, No. 2, pp. 160–168.

Nicolas, U., Tanguy, D., Hervé, G., & Jean, B. 2016. Impacts of Agricultural Land use Changes on Pesticide use in French Agriculture. European Journal of Agronomy, Vol. 80 pp. 34–40.

Nunes, C., & Auge, J.I. 1999. Land use and land cover change: Implementation strategy. In: IHDP report No. 10.

Openshaw, S. & Taylor, P. 1979. A Million or so Correlation Coefficients: Three Experiments on the Modifiable Areal Unit Problem. In: Wrigley, N. (ed.) Statistical Applications in the Spatial Sciences. Pion, London, pp. 127–144.

O'Sullivan, D. 2001. Graph-cellular Automata: A Generalised Discrete Urban and Regional Model. Environment and Planning B, Vol. 28, No. 5, pp. 687–706.

Ozcan, A.U., Erpul, G., Basaran, M., & Erdogan, H.E. 2008. Use of USLE/GIS Technology Integrated with Geostatistics to Assess Soil Erosion Risk in Different Land Uses of Indagi Mountain Pass—Cankırı, Turkey. Environmental Geology, Vol. 53, 8, pp. 1731–1741.

Ozsoy, G., Aksoy, E., Dirim, M.S., & Tumsavas, Z. 2012. Determination of Soil Erosion Risk in the Mustafakemalpasa River Basin, Turkey, Using the Revised Universal Soil Loss Equation, Geographic Information System, and Remote Sensing. Environmental Management, Vol. 50, No. 4, pp. 679–694.

Pace, R.K., Barry, R., & Sirmans, C.F. 1998, Spatial Statistics and Real Estate. Journal of Real Estate Finance and Economics, Vol. 17, pp. 5–13.

Paez, A., Uchida, T., & Miyamoto K. 2002. A General Framework for Estimation and Inference of Geographically Weighted Regression Models: Location-specific Kernel Bandwidths and a Test for Local Heterogeneity. Environmental and Planning A, Vol. 34, pp. 733–754.

Park, S., Oh, C., Jeon, S., Jung, H., & Choi, C. 2011. Soil Erosion Risk in Korean watersheds, Assessed Using the Revised Universal Soil Loss Equation. Journal of Hydrology, Vol. 399, No. 3–4, pp. 263–273.

Parker, D.C. 2000. Edge-effect Externalities: Theoretical and Empirical Implications of Spatial Heterogeneity. Ph D. diss. Davis, CA: University of California at Davis.

Parker, D.C., Manson, S.M., Janssen, M.A., Hoffmann, M.J., & Deadman, P. 2003. Multi-agent System Models for the Simulation of Land-use and Land-cover Change: A review. Annals of the Association of American Geographers, Vol. 93, No. 2, pp. 314–337.

Pavlov, A. 2000. Space Varying Regression Coefficients: A Semi-parametric Approach Applied to Real Estate Markets. Real Estate Economics, Vol. 28, pp. 249–283.

Peng, J., Liu, Y.X., Li, T.Y., & Wu, J.S. 2017. Regional Ecosystem Health Response to Rural Land use Change: A Case Study in Lijiang City, China. Ecological Indicators, Vol. 72 pp. 399–410.

Pontius, Jr., & Gilmore, R. 2000. Quantification Error Versus Location Error in Comparison of Categorical Maps. Photogrammetric Engineering & Remote Sensing, Vol. 66, No. 8 pp. 1011–1016.

Pontius, Jr., Gilmore, R., Walker, R., Yao-Kumah, R., Arima, E., Aldrich, S., Caldas, M., & Vergara, D. 2007. Accuracy Assessment for a Simulation Model of Amazonian Deforestation. Annals of the Association of American Geographers, Vol. 97, No. 4 pp. 677–695.

Portugali, J., Benenson, I., & Omer, I. 1997. Spatial Cognitive Dissonance and Sociospacial Emergence in a Self-organizing City. Environment and Planning B, Vol. 24 pp. 263–285.

Quétier, F., Lavorel, S., Thuiller, W., & Davies, I. 2007. Plant-trait-based Modeling Assessment of Ecosystem-service Sensitivity to Land-use Change. Ecological Applications, Vol. 17, No. 8 pp. 2377–2386.

Ramankutty, N. & Foley, J.A. 1999. Estimating Historical Changes in Global Land Cover: Croplands from 1700 to 1992. Global Biogeochemical Cycles, Vol. 13, No. 4, pp. 997–1027.

Rindfuss, R.R., Walsh, S.J., Turner II, B.L., Fox, J., & Mishra, V. 2004. Developing a Science of Land Change: Challenges and Methodological Issues. PNAS, Vol. 101, No. 39 pp 13976–13981.

Romero, H. & Ordenes, F. 2004. Emerging Urbanization in the Southern Andes - Environmental Impacts of Urban Sprawl in Santiago de Chile on the Andean Piedmont. Mountain Research and Development, Vol. 24, No. 3, pp. 197–201.

Saha, A., Gupta, R., & Arora, M. 2002. GIS-based Landslide Hazard Zonation in the Bhagirathi (Ganga) Valley, Himalayas. International Journal of Remote Sensing, Vol. 23, No. (2), pp. 357–369.

Sanders, L., Pumain, D., Mathian, H., Guérin-Pace, F., & Bura, S. 1997. SIMPOP - a Multiagent System for the Study of Urbanism. Environment and Planning B: Planning and Design, Vol. 24 pp. 287–305.

Sanderson, W.C. 1998. Knowledge can Improve Forecasts: A Review of Selected Socioeconomic Population Projection Models. Population and Development Review, Vol. 24 pp. 88–117.

Scott, A., Carter, C., Reed, M., Larkham, P., Adams, D., Morton, N., Waters, R., Collier, D., Crean, C., & Curzon, R. 2013. Disintegrated Development at the Rural–urban Fringe: Re-connecting Spatial Planning Theory and Practice. Progress in Planning, Vol. 83 pp. 1–52.

Seattle, S. 2006. Toward a Sustainable Future: The King County/Seattle Indicator & Strategies for Action Project.

Sharpley, A.N. & Williams, J.R. 1990. EPIC-erosion/productivity impact calculator: 1. Model documentation. Technical Bulletin-United States Department of Agriculture (1768 Pt 1).

Shenzhen Municipal Government, 2010. Shenzhen Urban Master Plan (2010–2020). Shenzhen. (In Chinese).

Shenzhen Municipal Government, 2011. Shenzhen Economic and Social Development Twelfth Five-Year Plan. (In Chinese).

Shenzhen Municipal People's Government, 2011. 12th Five Year Plan of National Economic and Social Development of Shenzhen.

Shenzhen Statistics Bureau, 2001. Shenzhen Statistical Yearbook. Beijing, National Statistics Press.

Shenzhen Statistics Bureau, 2003. Shenzhen Statistical Yearbook. Beijing, National Statistics Press.

Shenzhen Statistics Bureau, 2008. Shenzhen Statistical Yearbook. Beijing, National Statistics Press.

Shenzhen Statistics Bureau, 2009. Shenzhen Statistical Yearbook. Beijing, National Statistics Press.

Shenzhen Statistics Bureau, 2012. Shenzhen Statistical Yearbook. Beijing, National Statistics Press.

Shi, H., Zhang, L., & Liu, J. 2006. A New Spatial-attribute Weighting Function for Geographically Weighted Regression. NRC press.

Shi, P. & Yu, D. 2014. Assessing Urban Environmental Resources and Services of Shenzhen, China: A landscape-based approach for urban planning and sustainability. Landscape and Urban Planning.

Shrestha, S., Kazama, F., Newham, L., Babel, M., Clemente, R., Ishidaira, H., Nishida, K., & Sakamoto, Y. 2008. Catchment Scale Modelling of Point Source and Non-point Source Pollution Loads Using Pollutant Export Coefficients Determined from Long-term in-stream Monitoring Data. Journal of Hydro-environment Research, Vol. 2, No. 3, pp. 134–147.

Silberstein, J. & Maser, C. 2000. Land use Planning for Sustainable Development. CRC Press.

Silvertown, J., Holtier, S., Johnson, J., & Dale, P. 1992. Cellular Automaton Models of Interspecific Competition for Space: The Effect of Pattern on Process. Journal of Ecology, Vol. 80, No. 3 pp. 527–534.

Sirmans, G., Macpherson, D., & Zietz, A. 2005. The Composition of Hedonic Pricing Models. Journal of Real Estate Literature, Vol. 13, No. 1, pp. 3–41.

SmoyerTomic, K.E., Hewko, J.N., & Hodgson, M.J. 2004. Spatial Accessibility and Equity of Playgrounds in Edmonton, Canada. The Canadian Geographer/Le Géographe Canadien, Vol. 48, No. 3, pp. 287–302.

Sorkhabi, S.Y.D., Romero, D.A., Yan, G.A., Gu, M.D., & Moran, J. 2016. The Impact of Land use Constraints in Multi-objective Energy-noise Wind Farm Layout Optimization. Renewable Energy, Vol. 85, pp. 359–370.

Srivastava, P., Hamlett, J., Robillard, P., & Day, R. 2002. Watershed Optimization of Best Management Practices Using AnnAGNPS and a Genetic Algorithm. Water Resources Research, Vol. 38, No. 3, 3-1-3-14.

Standing Committee of the National People's Congress, 2007. Urban and Rural Planning Law of the People's Republic of China. Beijing.

Steuer, R.E. 1989. Multiple Criteria Optimization: Theory, Computation, and Application. Krieger Malabar.

Stewart, T.J., Janssen, R., & van Herwijnen, M. 2004. A Genetic Algorithm Approach to Multiobjective Land use Planning. Computers & Operations Research, Vol. 31, No. 14, pp. 2293–2313.

Stone, B., Hess, J.J., & Frumkin, H. 2010. Urban Form and Extreme Heat Events: Are Sprawling Cities More Vulnerable to Climate Change Than Compact Cities? Environmental Health Perspectives, Vol. 118, No. 10 pp. 1425–1428.

Talen, E. & Anselin, L. 1998. Assessing spatial equity: An evaluation of measures of accessibility to public playgrounds. Environment and Planning A, Vol. 30, No. 4 pp. 595–613.

Taleshi, M. & Ghobadi, A. 2012. Urban Land Use Sustainability Assessment Through Evaluation of Compatibility Matrix Case Study: Karaj City. OIDA International Journal of Sustainable Development, Vol. 3, No. 1 pp. 57–64.

Talbot, T.O., Kulldorff. M., Forand. S.P., & Haley. V.B. 2000. Evaluation of Spatial Filters to Create Smoothed Maps of Health Data. Statistics in Medicine, Vol. 19 pp. 2399–2408

Theobald, D.M. & Hobbs, N.T. 1998. Forecasting Rural Land Use Change: A Comparison of Regression and Spatial Transition-based Models. Geographical and Environmental Modeling, Vol. 2, pp. 65–82.

Tian, Y., Xiao, C.C., Liu, Y., & Wu, L. 2008. Effects of Raster Resolution on Landslide Susceptibility Mapping: A Case Study of Shenzhen. Science in China Series E: Technological Sciences, Vol. 51, No. 2 pp. 188–198.

Tianhong, L., Wenkai, L., & Zhenghan, Q. 2010. Variations in Ecosystem Service Value in Response to Land use Changes in Shenzhen. Ecological Economics, Vol. 69, No. 7 pp. 1427–1435.

Torrens, P.M. & O'Sullivan, D. 2001. Cellular Automata and Urban Simulation: Where do we go from here? Environment and Planning B, Vol. 28, No. 2 pp. 163–168.

Train, K. 2003. Discrete Choice Methods with Simulation. New York: Cambridge University Press.

Truelove, M. 1993. Measurement of Spatial Equity. Environment and planning C, Vol. 11, pp. 19–19.

Turner, B.L. & Meyer, W.B. 1991. Land Use and Land Cover in Global Environmental Change: Considerations for Study. International Social Sciences Journal, Vol. 130, pp. 667–679.

Turner, M.G., Costanza, R., & Sklar, F. 1989. Methods to Evaluate the Performance of Spatial Simulation Models. Ecological Modelling, Vol. 48, No. 1/2, pp. 1–18.

Upton, G.J.G. & Fingleton, B. 1985. Spatial Data Analysis by Example, Volume 1: Point Pattern and Quantitative Data New York: John Wiley and Sons.

Vaidya, O.S. & Kumar, S. 2006. Analytic Hierarchy Process: An Overview of Applications. European Journal of Operational Research, Vol. 169, No. 1, pp. 1–29.

Vanwalleghem, T., Gómez, J.A., Amate, J.I., Molina, M.G.T., & Vanderlinden, K. 2017. Impact of Historical Land Use and Soil Management Change on Soil Erosion and Agricultural Sustainability During the Anthropocene. Anthropocene.

Veldkamp, A. & Fresco, L.O. 1996. CLUE: A Conceptual Model to Study the Conversion of Land Use and its Effects. Ecological Modelling, Vol. 85, No. 2/3, pp. 253–270.

Veldkamp, A. & Lambin, E.F. 2001. Predicting Land-use Change. Agriculture, Ecosystems, and Environment, Vol. 85, No. 1–3, pp. 1–6.

Verburg, P.H., Soepboer, W., Veldkamp, A. Limpiada, R., Espaldon, V., & Sharifah Mastura, S. A. 2002. Modeling the Spatial Dynamics of Regional Land Use: The CLUE-S Model. Environmental Management, Vol. 30, No. 3, pp. 391–405.

Wang, P. 2006. Exploring Spatial Effects on Housing Price: The Case Study of the City of Calgary. Master Dissertation, University of Calgary, Canada.

Wang, H., Shen, Q., & Tang, B.S. 2015. GIS-based Framework for Supporting Land Use Planning in Urban Renewal: Case Study in Hong Kong. Journal of Urban Planning and Development, Vol. 141, No. 3.

Weinberg, M., Kling, C.L., & Wilen, J.E. 1993. Water Markets and Water Quality. American Journal of Agricultural Economics, Vol. 75, No. 2, pp. 278–291.

Weng, Q. 2002. Land use Change Analysis in the Zhujiang Delta of China Using Satellite Remote Sensing, GIS, and Stochastic Modeling. Journal of Environmental Management, Vol. 64, pp. 273–284.

Wierzbicki, A.P. 1980. The Use of Reference Objectives in Multiobjective Optimization. Multiple Criteria Decision Making Theory and Application. Springer. pp. 468–486.

Wischmeier, W.H. & Smith, D.D. 1978. Predicting Rainfall Erosion Losses-a Guide to Conservation Planning. Predicting Rainfall Erosion Losses – A Guide to Conservation Planning.

Wolter, P.T., Johnston, C.A., & Niemi, G.J. 2015. Land Use Land Cover Change in the U.S. Great Lakes Basin 1992 to 2001. Journal of Great Lakes Research, Vol. 32, No. 3, pp. 607–628.

Wooldrige, J.M. 2002. Econometric Analysis of Cross Section and Panel Data. Cambridge, MIT Press.

Worrall, F. & Burt, T. 1999. The Impact of Land-use Change on Water Quality at the Catchment Scale: The Use of Export Coefficient and Structural Models. Journal of Hydrology, Vol. 221, No. 1, pp. 75–90.

Wu, F. & Yeh, A.G. 1997, Changing Spatial Distribution and Determinants of Land Development in Chinese Cities in the Transition from a Centrally Planned Economy to a Socialist Market Economy: A Case Study of Guangzhou. Urban Studies, Vol. 34, pp. 1851–1879.

Wu, F. 1998, SimLand: A Prototype to Simulate Land Conversion through the Integrated GIS and CA with AHP-derived Transition Rules. International Journal of Geographical Information Science, Vol. 12, pp. 63–82.

Xie, G., Lu, C., Leng, Y., Zheng, D., & Li, S. 2003. Ecological Assets Valuation of the Tibetan Plateau. Journal of Natural Resources, Vol. 18, No. 2, pp. 189–196.

Xie, M., Wang, Y., Fu, M., & Zhang, D. 2013. Pattern Dynamics of Thermal-environment Effect during Urbanization: A Case Study in Shenzhen City, China. Chinese Geographical Science, Vol. 23, No. 1, pp. 101–112.

Xu, Y. & Li, T. 1999. Dynamic Evaluation of Soil Erosion in Shenzhen Based on Land-use Structural Variation. Journal of Yangtze River Scientific Research Institute, Vol. 26, No. 7, pp. 6–12.

Yamada, I., Brown, B.B., Smith, K.R., Zick, C.D., Kowaleski-Jones, L., & Fan, J.X. 2012. Mixed Land use and Obesity: An Empirical Comparison of Alternative Land use Measures and Geographic Scales. The Professional Geographer, Vol. 64, No. 2, pp. 157–177.

Yanefski, J. 2013. Land Use Profile of China. http://www.eoearth.org/view/article/154148.

Yang, D., Kanae, S., Oki, T., Koike, T., & Musiake, K. 2003. Global Potential Soil Erosion with Reference to Land Use and Climate Changes. Hydrological Processes, Vol. 17, No. 14, pp. 2913–2928.

Yang, J., Wang, C.Q., Xia, J.G., Zhang, W.X., Bai, G.C., & Zhang, Y. 2010. Spatial Planning of Land use Based on Cellular Automata Modeling: A Case Study of Dongpo District, Meishan City. Acta Pedologica Sinica, Vol. 47, No. 5, pp. 847–858. (In Chinese).

Yang, L., Hao, J.M., Lv, Z.Y., Chen, H.Y., & Wang, Y. 2012. Analysis and Prediction of Carbon Sources and Sinks in Quzhou County of Huang Huaihai Plain, China. Journal of Food, Agriculture & Environment, Vol. 10, No. 1, pp. 656–661.

Yang, Z., Zhu, X., & Moodie, D.R. 2014. Optimization of Land Use in a New Urban District. Journal of Urban Planning & Development, Vol. 141, No. 2, 0501 4010.

Yao, X., Wang, Z., & Wang, H. 2015. Impact of Urbanization and Land-Use Change on Surface Climate in Middle and Lower Rreaches of the Yangtze River, 1988–2008. Advances in Meteorology, No. 4, pp. 1–10.

Ye, L.M. & Van Ranst, E. 2002. Population Carrying Capacity and Sustainable Agricultural use of Land Resources in Caoxian County (North China). Journal of Sustainable Agriculture, Vol. 19, No. 4 pp. 75–94.

Yin, Y.Y. & Pierce, J.T. 1993. Integrated Resource Assessment and Sustainable Land-use. Environmental Management, Vol. 17, No. 3, pp. 319–327.

Young, W.J., Marston, F.M., & Davis, R.J. 1996. Nutrient Exports and Land use in Australian Catchments. Journal of Environmental Management, Vol. 47, No. 2, pp. 165–183.

Yu, D-L. 2006. Spatially Varying Development Mechanisms in the Greater Beijing Area: A geographically Weighted Regression Investigation. The Annals of Regional Science, Vol. 40, Vol. 1, pp. 173–190.

Yuan, M. & Stewart, K. 2008. Computation and Visualization for the Understanding of Dynamics in Geographic Domains: A Research Agenda. CRC/Taylor and Francis.

Zhang, H.H., Zeng, Y.N., & Bian, L. 2010. Simulating Multi-objective Spatial Optimization Allocation of Land use Based on the Integration of Multi-agent System and Genetic Algorithm. International Journal of Environmental Research, Vol. 4, No. 4, pp. 765–776.

Zhang, H.H., Zeng, Y.N., & Bian, L. 2011. Multi-objective Spatial Optimization Model for Land use Allocation and its Application. Journal of Central South University (Science and Technology), Vol. 4, pp. 034.

Zhang, Q. 2007. Study on Water Ecology Security Planning in Shenzhen. Beijing University of Chemical Technology.

Zhang, X.P. & Cheng, X.M. 2009. Energy Consumption, Carbon Emissions, and Economic Growth in China. Ecological Economics, Vol. 68, No. 10, pp. 2706–2712.

Zhang, X.Q., Yang, S.J., Xiang, W.N., & Wang, A.P. 2007. Spatial Planning Method for the Basic Farmland Protection Based on the Farmland Classification. Transactions of the CSAE, Vol. 23, No. 1, pp. 66–75. (In Chinese).

Zhang, Z. 2000. Decoupling China's Carbon Emissions Increase from Economic Growth: An Economic Analysis and Policy Implications. World Development, Vol. 28, No. 4, pp. 739–752.

Zhao, B., Kreuter, U., Li, B., Ma, Z., Chen, J., & Nakagoshi, N. 2004. An Ecosystem Service Value Assessment of Land-use Change on Chongming Island, China. Land Use Policy, Vol. 21, No. 2 pp. 139–148.

Zhao, Y.X., Zhang, W.S., Wang, Y., & Wang, T.T. 2007. Soil Erosion Intensity Prediction Based on 3S Technology and the USLE: A Case from Qiankeng Reservoir Basin in Shenzhen. Journal of Subtropical Resources and Environment, Vol. 3, 006.

Zhou, W.Z., Xie, S.Y., Zhu, L.F., Tian, Y.Z., & Yi, J. 2008. Evaluation of Soil Erosion Risk in Karst Regions of Chongqing, China. Optical Engineering Applications: 70831E-70831E-70810.

Zhu, J., Huang, H., & Wu, J. 2005. Modelling Spatial-Temporal Binary Data Using Markov Random Fields. Journal of Agricultural, Biological & Environmental Statistics, Vol. 10, No. 2, pp. 212–216

Zhu, L.W. 2010. Literature Review of Land use Planning. Economic & Trade Update, Vol. 18, No. 26, pp. 112–114. (In Chinese).

Zobrist, J. & Reichert, P. 2006. Bayesian Estimation of Export Coefficients from Diffuse and Point Sources in Swiss watersheds. Journal of Hydrology, Vol. 329, No. 1, pp. 207–223.

Printed and bound by CPI Group (UK) Ltd, Croydon, CR0 4YY

01/11/2024

01782604-0004